The Evolution of American
Women's Studies

Previous Publications by Alice E. Ginsberg

Alice E. Ginsberg and Marybeth Gasman (Eds.) *Gender and Educational Philanthropy: New Perspectives on Funding, Collaboration, and Assessment* (2007).

Alice E. Ginsberg, Joan Poliner Shapiro, and Shirley Brown, *Gender in Urban Education: Strategies for Student Achievement* (2004).

The Evolution of American Women's Studies

Reflections on Triumphs, Controversies, and Change

Edited by Alice E. Ginsberg

THE EVOLUTION OF AMERICAN WOMEN'S STUDIES
Copyright © Alice E. Ginsberg, 2008.

First published in 2008 by
PALGRAVE MACMILLAN®
in the United States—a division of St. Martin's Press LLC,
175 Fifth Avenue, New York, NY 10010.

Where this book is distributed in the UK, Europe and the rest of the world,
this is by Palgrave Macmillan, a division of Macmillan Publishers Limited,
registered in England, company number 785998, of Houndmills,
Basingstoke, Hampshire RG21 6XS.

Palgrave Macmillan is the global academic imprint of the above companies
and has companies and representatives throughout the world.

Palgrave® and Macmillan® are registered trademarks in the United States,
the United Kingdom, Europe and other countries.

ISBN-13: 978–0–230–60579–4
ISBN-10: 0–230–60579–6

Library of Congress Cataloging-in-Publication Data

The evolution of American women's studies : reflections on triumphs,
controversies, and change / edited by Alice E. Ginsberg.
    p. cm.
ISBN 0–230–60579–6
    1. Women's studies—History. 2. Feminism. I. Ginsberg, Alice E.

HQ1180.E86 2008
305.40973—dc22   2008016033

A catalogue record of the book is available from the British Library.

Design by Newgen Imaging Systems (P) Ltd., Chennai, India.

First edition: November 2008

10 9 8 7 6 5 4 3 2 1

Printed in the United States of America.

This book is dedicated in memory of my mother
Lois Taterka Ginsberg
1937–1907

Contents

Foreword

I am honored to write this forward to Alice Ginsberg's important collection *The Evolution of American Women's Studies: Reflections on Triumphs, Controversies, and Change* because I am a product of the hard work many of these contributors have done to institutionalize women's studies in the academy, and am literally and figuratively a student and colleague of many writers featured here. I hold a Ph.D. in Women's Studies from Emory University, and was among the first wave of students earning the Women's Studies Ph.D. in the United States. I was a graduate student when much scholarly debate took place about the validity, rigor, and merits of women's studies doctoral education, as Ginsberg notes in Part One. Though it was unsettling to be pursuing doctoral study in a field whose merits and value were highly contested by some, it was also exhilarating and pioneering work; my peers and I were helping to build a graduate program and advance a field. Our experience led us to make suggestions about curricular changes and program requirements that were subsequently implemented: the power of student voices to influence change should come as no surprise, because this is part of the women's studies tradition, as many contributors note here.

For me, women's studies has always been about the intersectional analysis of multiple identity categories due in no small part to taking a course on Black Women in Historical Perspective taught by Beverly Guy-Sheftall, a contributor to this collection and a pioneering women's studies scholar-activist herself. Unlike many other faculty members at Emory who came to women's studies from academic homes in traditional disciplines, Guy-Sheftall saw herself as a women's studies scholar first and foremost. I have since heard Beverly talk about how being in women's studies at Spelman, a historically black women's college, gave her a unique perspective on her place in the field; she explained she always understood that she had a place at its center. Looking back, I am certain she taught her course as an Emory visiting faculty member from that epistemological position: of course Black women's theories, histories, knowledge, and struggles were central to the work of the field, and of course doctoral students in women's studies would need to understand those theories, histories, knowledge, and struggles.

I am grateful for that training and have applied it to my work at the National Women's Studies Association, where I now serve as its executive director. Here, too, I take my place as part of a long tradition of scholarship and activism; anecdotes about NWSA's founding and conferences run throughout these essays. As Ann Russo notes here, making women of color knowledge and perspectives central to the work of the Association and the field is a critical, ongoing project undertaken by many throughout the organization. As Russo indicates, NWSA—like the feminist movement and women's studies more broadly—must constantly work to overcome its history of racism and white supremacy.

While we are all the products of our shared history of racism, our work in NWSA and in the field of women's studies is also inevitably forward-looking, because we are committed to "educational and social transformation," as the current NWSA mission statement indicates. Indeed, every scholar included in this volume speaks to the potential for change that women's studies imagines. NWSA recently completed a strategic planning process that is designed to position the organization as leading the field through its next stages of growth, change, and intellectual debate. One important part of that planning process included collecting data on the field of women's studies nationally, because to imagine where we want to go, we also have to take stock of where we are. The data collection project was intended to provide quantitative information that would help make the case for the field and resources to support women's studies work. Perhaps not surprisingly, we have learned that women's studies programs are more racially diverse than faculty nationally, that women's studies courses typically fulfill general education requirements, and that many women's studies programs operate with modest or nonexistent budgets.

After reading the essays here, I am inclined to view our data as yet another "both/and" portrait of women's studies not unlike the ones offered in this volume. Women's studies faculty may no longer have to meet in the "ladies" room or be consigned to "special" academic units as some contributors recall here, but in many ways women's studies still operates at institutional margins with limited resources. And yet, the field certainly made progress toward becoming institutionalized and garnering respect, with strong growth at the doctoral level. These tensions reflect what is, in my view, most powerful about women's studies: a creative space in which new visions for the future can emerge.

ALLISON KIMMICH
Executive Director, National Women's
Studies Association (NWSA)

Reference

"A National Census of Women's and Gender Studies Programs," National Women's Studies Association, 2007.

Acknowledgments

I first became interested in feminism and women's liberation in high school, and, as I describe later in the Introduction, decided to become a women's studies major my first semester at college during my first *Introduction to Women's Studies* course. It is with great excitement that I publish this book, as I have always felt great pride in being a women's studies major and the subsequent work I have done—almost all in the area of women's studies and gender studies.

Most importantly, I am thrilled at the quality of the contributors to this volume. First and foremost, I thank them all.

I could never have done any of this work, without the encouragement and support of many different people in my life. I want to thank my editors at Palgrave, Julia Cohen, Brigitte Shull, Kristy Lilas, and Amanda Moon; the team at Newgen; and my colleagues and coauthors of previous works Shirley Brown, Joan Poliner Shapiro, and Marybeth Gasman all of whom continue to inspire me on an ongoing basis. I would like to thank all my women's studies teachers for mentoring me and showing me why women's studies was such a unique endeavor and why gender was such an important category of analysis. I would also like to thank Allison Kimmich, Executive Director of the National Women's Studies Association (NWSA), for writing the Foreword.

I want to thank my good friends and close family—all of whom have helped me through a very tough year in many different regards. Their love and multifaceted support was, and continues to be, critical to me as I move forward.

I want to acknowledge my sons, Andrew and Nicholas Chalfen, of whom I could not be more proud of or pleased with. It is a great joy to watch them grow and learn and to see what wonderful ideas they have already at ages nine and thirteen.

Of course, I want to thank my husband, Samuel Chalfen, who after 19 years is still my best friend and most cherished partner. As with my past two books, and all my writing, Sam has always given me important support, feedback, and encouragement. He has also been there for me through thick and thin.

Finally, I want to thank my parents, Lois and Ralph Ginsberg. They deserve special thanks, as without really even knowing what women's studies was, they supported my decision to major in something that wasn't widely respected and didn't have a clear "career" attached to it. I especially want to thank my mom—to whom this book is dedicated—for reading drafts of all my articles, proposals, reports, manuscripts, and the like, and always giving me feedback that enhanced these writings for the better. In the process, I think a little bit of my overzealousness for feminism rubbed off on her, although she didn't really need it, as she was one of the strongest, most productive, and most caring people I ever knew.

Introduction

The Evolution of American Women's Studies: Reflections on Triumphs, Controversies, and Change

Alice E. Ginsberg

Evolutions

This book explores the evolutions of American Women's Studies from its first entry into the academy in the 1960s and 1970s to present-day politics and programs. The changes, as will be illustrated, are dramatic at the same time that they naturally mirror larger changes in both American and academic politics, culture, and history. It has been said that when women' studies first entered the university "merely to assert that woman should be studied was a radical act" (Boxer, 2001, 10). Nearly 40 years later, there are over 800 women's studies programs nationwide, including opportunities for students to get their B.A., graduate certificate, M.A., and even Ph.D. in the field (Mohanty, 2005, 76). Moreover, the focus of Women's Studies has blossomed to include significant research on girls *and boys* in elementary and secondary school settings.

Over these years, every aspect of the field has been questioned and debated, ranging from 1) whether Women's Studies should be thought of as an academic "discipline" (given it's inherently interdisciplinary nature), 2) whether Women's Studies should be called Women's Studies (e.g., as opposed to Feminist Studies, Female Studies, Gender Studies, Sexuality Studies, etc.), 3) whether Women's Studies adequately addresses the experiences of a diversity of women, and

4) how Women's Studies has developed alongside of and interacted with other interdisciplinary "identity-based programs" such as American Studies, African American Studies, Jewish Studies, and the like.

These are not new questions, and there are many good books on the theory and history of feminism and women's studies. So why another? Typically, books on Women's Studies tend to be broken up into chapters on the impact of Women's Studies in different disciplines, such as women's literature, women's history, or else they are designed to individually explore the experiences of women of different races, classes, and sexual identities. For example, one chapter may be on African American women's experiences, while another is focused on lesbian woman's experiences. While this is not a problem per se, these books often make it difficult for the reader to see the "larger picture" of the evolution of feminism and Women's Studies *across* time, disciplines, and identities.

Moreover, while many such books, or chapters therein, are written by Women's Studies professors and practitioners, they are often very "academic" in style and tone, taking for granted that the reader knows about "French Feminism," "modernism," or "post-structuralism," which has the potential to alienate many people who are interested in the field but not schooled in such theory.

By contrast, this book opens with a brief historical essay exploring evolutions in Women's Studies over the past 40 years as they occurred *chronologically*, and the majority of the book is primarily comprised of *reflective essays* written by influential Women's Studies scholars. These essays address their actual experiences in the feminist movement and in the Women's Studies classroom over a period of years and through a range of political debates about the very nature of feminism and Women's Studies. They question Women's Studies' right to exist, as well as where it stands within the university disciplinary structure. They look at what should be its primary goals (e.g., should academic scholarship also have the aspiration of social change?). They consider whether or not the field should be *for* women (e.g., the apostrophe in women's) *about* women, or about gender, suggesting, among other things, that boys and men need to be an integral part of Women's Studies theory, research, and analysis.

In writing these essays the scholars were asked to consider a series of questions, including:

- Why were you interested in getting involved in Women's Studies? Did you have any personal experiences or mentors that facilitated this process?

- How has your definition of Women's Studies' goals and purposes changed over the years?
- Do you define Women's Studies as a radical or subversive act? Why/Why not?
- What is the relationship between Women's Studies and women's liberation?
- Do you think Women's Studies should be its own discipline despite its inherently interdisciplinary nature? If not, how should we define and organize it?
- Have the demographics and views of your students changed considerably over the years and, if so, in what ways?
- What kinds of "texts" have you used over the years and what kinds of texts are you using now (e.g., Literature in the field has proliferated greatly. How have you made use of this?)
- How has your work addressed differences between and among women? How are these issues raised in your classes and what has been the spectrum of responses from students?
- Likewise, how have you worked with other related academic departments, such as African American studies, and so on?
- What kinds of Women's Studies networks have you been involved in and what have been the purposes and outcomes of these networks?
- What political issues have been most prominent across the decades that you have been teaching?
- What should Women's Studies be called (e.g., gender studies, feminist studies, etc.)?
- Overall, how has your thinking about Women's Studies as a field within academia changed during the time you have been involved in it?
- What was the biggest political issue that you faced?
- Where do you see Women's Studies in 30 years?

The ways in which these questions are answered vary considerably, as I have specifically chosen a group of Women's Studies scholars who are diverse in multiple ways, including: how long they have been in the field; what disciplines they were trained in and/or are considered experts in; what kind of academic institution they work in and where it is located geographically; whether they work primarily in the classroom or in a more administrative programmatic position; and whether their work focuses on women or *girls* in school and society. Perhaps most importantly, this group of scholars includes women of many different races, classes, ethnicities, religious and sexual identities. Thus, when we speak of the "evolution" of Women's Studies, it is important to emphasize that this is a multifaceted process. Women's Studies has differed widely over time, and across identities, disciplines, and institutions. The contributions to this book will highlight those differences. They also write from a personal experience, so many tell the stories of their own "evolutions."

My Own Experience

Later/ A Haiku for Jennifer
Suddenly she mourns
Her unborn baby
In our anonymous arms

I wrote this Haiku in 1981 while taking my first women's studies course. The class was taught by Professor Sonia Sanchez, a well-known and very talented African American feminist, author, and poet, who was an incredibly dynamic and moving teacher. We were having a discussion on writer Gwendolyn Brooks' story "The Mother," and one student suddenly burst into tears. Although she did not talk at great length as to what she had gone through in deciding whether or not to get an abortion, the issue quickly shifted from one of pure "intellectual discussion" to a recognition that has been said so many times about women's studies, that the "personal is political."

I have thought about this incident a number of times over the 20 plus years since I have graduated from college because it was so atypical compared to the rest of my education. It was in this class, my very first semester of college, that I decided I wanted to be a women's studies major. I was drawn to the discipline not only because it connected theory with practice, but because it challenged us to ask questions and explore each other's experiences and points of view, rather than to simply memorize information and reiterate it. In short, we functioned as a community of learners rather than as a group of students who just happened to sit next to each other.

I include this here because I think it appropriate to say something about my own experience. I got my B.A. in women's studies in 1985 from Temple University, where I was one of the first generations of women to have that option. I recently found a journal from my college days in which I wrote that for me, women's studies was as much about "studies" as about women. What I meant by that was that in addition to caring about women's equality and quality of life, I was very interested in the pedagogy of women's studies, which contrasted so sharply with my other courses of learning. Rather than assuming that knowledge was preexisting and above question, in my women's studies classes we actually participated in gathering and creating new knowledge, a significant amount of which had been overlooked and undervalued in academic life since its inception.

I liked the fact that a question could be answered from many different perspectives and still be "right." Women's Studies was also very "student centered" in it's approach (e.g., as opposed to a lecture/teacher as expert model); it valued personal experience as the basis for understanding larger political issues; it was interdisciplinary in nature; and perhaps most importantly, it stressed that our theories should be grounded in our actions and that part of the value of women's studies in the academy was to make a real and profound difference in the lives of women *outside of* the academy—and not just locally, but internationally as well.

Perhaps more importantly, rather than looking at women through a narrow lens where women of privilege, like men of privilege were the "norm," issues of race, class, ethnicity, religion, sexuality, and such were always present, causing us to question our assumptions about everything, including what defined the category of woman itself.

As will be described in the chapter one, addressing differences among and between women, and acknowledging that women can actually participate in the oppression and silencing of other women, is still a difficult issue in the field. Nonetheless, even in 1985 this issue was on the table and, for me, made the field that much more challenging and exciting. We weren't just studying women, we were questioning the very binary oppositions (e.g., man/woman) that defined them.

Although Women's Studies had been in the academy for over 15 years at the time I got my B.A., many people had still never even heard of it. Far more people asked me at the time what I was going to "do" with a major in Women's Studies, a question I answered publicly and proudly in an article in the *Temple Review*, excerpted herein:

To ask "What are you going to do with it?" implies that education is a passive process. It implies that we learn and then we do. But in many ways the very nature of women's studies, which grew out of and alongside the women's liberation movement, is attractive because it is already active. Women's studies was born out of the political realities of women's lives. Women's studies was born not just out of the desire to learn, but the necessity to learn; not so much out of literature, as out of illiteracy, not so much out of economics as out of poverty....

I learned to look at a problem from many different perspectives knowing that there didn't always have to be a right or wrong way to approach something. I learned the value of working cooperatively, that everybody has a voice, and should be given an opportunity to use it. I learned that my experience as a white, middle-class woman was not at the center of history and that history itself needed to be rewritten to include other cultures, races and ethnicities. I learned that theory and practice should go hand in hand. I learned that education should be about change and evolution, and not about reiterating what is already known. I take that knowledge with me to each job and do with it—whatever I can. (Ginsberg, 1992)

Unfortunately, I suspect that this question, "what are you going to do with it?" is still being asked of students today, despite the fact that almost every college and university has a Women's Studies program or department and issues of gender equity in education, starting at pre-K, have made it to the mainstream media (e.g., discussed on Oprah, etc.)

It is my hope that the reflections in this book not only shed new light on Women's Studies accomplishments and challenges, but encourage its readers to ask questions of their own about the value and future of the field, their role within it, and their connection to the larger feminist movement. Even as this book is being published, women's studies is changing and expanding. As Zimmerman has rightly noted: "Women's Studies truly is and always will be a field in motion" (Zimmerman, 2005).

References

Boxer, M. (2001). *When women ask the questions: Creating Women's Studies in America*. Washington, DC: Johns Hopkins University Press.

Ginsberg, A.E. (1992). "What one *does* with a major in Women's Studies." *Temple Review* (Winter): 47.

Mohanty, C.T. (2005). "Under Western eyes" Revisited: "Feminist solidarity through anticapitalist struggles." In E. Kennedy and A. Beins (Eds.), *Women's Studies for the future: Foundations, interrogations, politics*. New Jersey: Rutgers University Press, 72–96.

Zimmerman, B. (2005). "Beyond dualisms: Some thoughts about the future of Women's Studies." In E. Kennedy and A. Beins (Eds.), *Women's Studies for the future: Foundations, nterrogations, olitics*. New Jersey: Rutgers University Press, 31–39.

Part One

History

Triumphs, Controversies, and Change: Women's Studies 1970s to the Twenty-First Century

Alice E. Ginsberg

Can we conceive of a future in which oppositional gender categories are not fundamental to one's self-concept?

(Alcoff, 1988, 287)

This chapter, which provides a historical and chronological overview of the evolution of women's studies over the past 40 years, is meant to provide context for the reflections that follow. Divided by each of the decades, it is not meant to be an exhaustive history of women's studies but rather a "snapshot" of how the field evolved as it interacted with issues in the women's liberation movement, American politics, and social change, as well as in the academy itself.

It is important to note, however, that changes in the field did not fall "neatly" into the decades, nor did different programs evolve in the same way at the same time. Many significant questions—such as: whether women's studies is or should, in fact, be its own discipline; whether women's studies pays adequate attention to differences among and between women; how closely women's studies should be tied to feminism and the active quest for women's equality *outside* of the academy; what women's studies should be called; what is the role of males in women's studies, and so on—were present from the beginnings of academic feminism and are still present today. These are pervasive questions in the field. Thus, many of these issues are brought up multiple times within different decades as they faced different *triumphs, controversies,* and/or *change.*

Women's Studies in the 1960's/1970's

> When Women's Studies was born in the mid-1970s, politics was its midwife. (Jean Robinson, 2002, 202)

> The first public meeting of my women's liberation group in the mid-1960's was held in the "Ladies Room" of the Pennsylvania State University (PSU) student union building.... When the university denied us a meeting room, we marched down the hall, crossed out "Ladies" put up "Women," propped open the door, and proceeded to have our meeting in the outer restroom area. (Farley, 2000, 265)

Women's Studies courses began to appear in colleges and universities in the late 1960s. The first official women's studies *program* was at San Diego State University in 1970. As the first quote above suggests, women's studies grew naturally out of the politics of its time—mainly the women's liberation movement, the Civil Rights movement, the movement for gay and lesbian equality, and the protests against Vietnam War. From its very inception, Women's Studies had a very clear purpose and that was to transform the university so that knowledge about women was no longer invisible, marginalized, or made "other."

It was also explicit within the founding that studying women in an academic setting would directly help make positive social change in the rest of the world for women and other oppressed groups alike. In other words, theory and practice were to go hand in hand. As Marilyn Boxer (2001) has noted "From the beginning, the goal of women's studies was not merely to study women's position in the world but to change it" (13). As likewise noted by a women's studies student in Caryn McTighe Musil's *The Courage to Question* (1992), a collection of case studies of women's studies programs across the country:

> That's the big difference in women's studies [from other disciplines]: there's not only the opportunity to argue, but there's almost a challenge to do something about it. (117)

Finally, as opposed to the title Female Studies, which was popular in the early foundations of the field, the advent of the title *Women's Studies* was significant in that "the apostrophe blurred the differences between studies by, about and belonging to women" (Boxer, 2001, 13). For the first time, women were not only learning about themselves, but were actively *creating* and *owning* knowledge based on their own personal and political experiences.

As the second quote above suggests, however, Women's Studies was not welcomed into the university with open arms. It was not unusual for programs to form around meetings in bathrooms and broom closets. Myra Dinnerstein (2000) recalls:

> During the three years we fought for a program we accrued an education, not only in feminist politics, but in university politics, becoming aware that understanding how to make our way through the system would be a necessary prerequisite to our success, not only in establishing the program but in its continued success. (2000, 294)

In the 1970s many women's studies courses and events were advertised through flyers, mimeographed newsletters, and by word of mouth. Weigman (2002) has noted that when women's studies first began to bloom in the early 1970s, feminism in the U.S. academy was not so much an "organized entity" as a group of courses, many of them listed on bulletin boards and/or taught for free. Also, in the early years, there were rarely funds for administrative/staff positions— much less full-time and/or tenure track appointments. In fact, many Women's Studies scholars worried that their association with the field would reflect poorly on them and derail their careers in their various disciplines (e.g., they would not get tenure).

It has also been said that when women' studies first entered the university "merely to assert that woman should be studied was a radical act" (Boxer, 2001, 10). It has noted that since most of those teaching women's studies classes were appointed in other disciplines, women professors did this work mostly on their own time, as an additional set of tasks to what they had been hired to do.

It was also difficult for scholars trained in a traditional discipline to do the interdisciplinary work required of women's studies. For example, a historian who specializes in women's history might have felt very uneasy teaching about new investigations into women's psychology or the sciences as a women's studies introductory class or as a women's studies theory course might require. (As I explore later in this essay, the emergence of women's studies graduate work, particularly the Ph.D., may change this dynamic, as a new generation of scholars will be trained in a new kind of "discipline.")

Moreover, those who are brave enough to venture outside their areas of specialization are often looked upon, as Aiken et al. (1988) have remarked:

> ...at best as neophytes—stereotypical tourists who cannot speak the language of the field they presume to examine and who overlook nuances and complexities apparent to the natives; at worst as

dangerous trespassers, or colonizers seeking to explore territory of
their own. (142–143)

Aiken et al. (1988) further note that "the language of 'scholarship'
is inseparable from the language of power," that is, "who owns the
discourse becomes inseparable from questions of ownership of all
sorts" (143).

Despite these challenges, students and faculty alike were determined to introduce this new area of study into the academy. And
this involved more than simply adding courses with "women" in the
title, or token readings by and about women in the syllabus. It meant
rethinking the very structure and foundations on which these disciplines were built. This meant asking new questions and providing
answers from a variety of perspectives, identifying new sources of
information, using new research methods and pedagogical practices,
new categories of analysis, and new means of evaluation. In other
words, it was not just a question of *what* was taught, but *how* and
why it was taught. How and why were women and other marginalized groups excluded in the first place? How must we rethink the
dominant assumptions of the field if we no longer predicate it on male
history and perspectives? When we genuinely address issues of gender, race, class, sexuality, and the like, in what ways must we also
change our theoretical assumptions?

Indeed, many students of Women's Studies took courses focused on
women without the benefit of an Introduction to Women's Studies course
or a feminist theory class. During 1971–1972, Feminist Press founder
Florence Howe and Carol Ahlum researched the state of women's studies and found few similarities among the more than 800 courses they
reviewed. As late as 1978, Howe called women's studies an "eclectic
smorgasbord" (1997, 29). Du Bois (1983) has similarly suggested that
taking courses from many different disciplines is likened to a jigsaw puz-
zle in which one believes that eventually they will see the whole picture.
Yet creating Women's Studies has been as much about *the process and
purpose of putting the pieces* together as it is about the resulting image.

Perhaps summed up most simply, "A syllabus is not an agenda"
(Boxer, 2001, 220). Women's Studies needed to define its mission with
the academy without alienating the critically important work being
done toward women's equality *outside* the academy. It also needed to
secure a place for itself within the academy where it was not dependent
on the "charity" of other departments to offer courses with women in
the title. In short, it needed to claim its own agenda.

Today, most women's studies programs offer degrees in women's studies and have Web sites with mission statements. Yet in the 1970s, it was still not entirely clear to those outside the field as to what women's studies was and what it was trying to accomplish (ranging from home economics to a radical feminist overthrow of the government and rejection of the American family). The overt "politics" of feminist theory and pedagogy stood in stark contrast to the idea that academic knowledge was purely "objective." Women's Studies was viewed with great suspicion, which would increase significantly as it came of age in the 1980s.

Later in this essay, and throughout the reflections that follow, the question of whether or not Women's Studies should be its own discipline—despite its inherently interdisciplinary nature—will be addressed in greater detail. As we jump from its early beginnings in the 1970s to present day, this question has still not been resolved. Indeed, one of the potential contributors to this volume immediately questioned its working title *Women's Studies: The Evolution of a Discipline*, noting that there is still no agreement on the subject of whether women's studies is, in fact, a discipline in its own right. This very question was the focus of the first meeting of the National Women's Studies Association (NWSA) in 1979, suggesting that it has not been an easy one to answer.

Many Women's Studies programs and feminist scholars have resisted the term "discipline" both theoretically and practically. Indeed, I've seen Women's Studies referred to as multidisciplinary, intradisciplinary, nondisciplinary, antidisciplinary, neo-disciplinary, transdisciplinary, cross- disciplinary, critical interdisciplinary, intersectional, intertexual, and pluri-disciplinary. I've also seen Women's Studies referred to as an "interdisciplinary discipline," and "interdiscipline." The question of how to define Women's Studies remains, and indeed has become even more problematic with the advent of Women's Studies graduate degrees. (Again, this will be discussed in more detail later in this chapter.)

Yet in the 1970s, the challenge was not so much whether to create a new discipline, but how to get Women's Studies recognized and funded by the university, period. As more and more colleges and universities embraced Women's Studies, they were faced with a lack of meeting space, administrative and academic staff, and opportunities for conferences and sharing of syllabi and pedagogy. There were few opportunities to share work across programs in different academic settings, and it was not uncommon to find academics reinventing the wheel as they struggled to gain legitimacy in their individual settings. Moreover, many serious Women's Studies scholars worried that the field was a "fad" and that, as Gloria Bowles (2000) has reflected:

> In 1978 I initiated a new seminar, Theories of Women's Studies. This was the thinking behind it: I had regarded women's studies as a temporary pursuit. We'd all go back to our disciplines eventually. (152)

The forming of a Women's Studies community was critical because despite the fact that these programs had many different histories and designs, Women's Studies programs have faced many similar questions and debates, as they closely paralleled historical changes in education and politics across the nation. I believe the word "evolution" is an important word because it suggests that Women's Studies did not exist in a vacuum, but closely paralleled changes throughout social and academic politics. It also suggests that programs influenced and learned from each other. During this time the field grew exponentially with the development of multiple women's research centers, special interest groups (SIGS) within larger educational organizations, and its own national association (NWSA)—a very important contributor of resources, conferences, and networking. There were also numerous new important journals such as *SIGNS, Feminist Frontiers, Feminist Teacher, Women's Studies Quarterly, SAGE*, and, more recently, online list-serve discussions (also to be discussed later in this chapter).

Women's Studies began and persists today mainly due to the hard work of students, teachers, administrators, funders, and other key stakeholders who strive to keep knowledge about women from being marginalized or invisible. Indeed, Nancy Topping Brazin (2000) recalls in the early days of the field that:

> Female graduate students in the English department took their own initiative in creating a women writers course called Literature and the Feminist Imagination. They designed the course, selected the texts, and asked me to teach it. Although the graduate advisor actively discouraged students from signing up for it, the course filled with thirty students. (62)

Nonetheless, women's studies was and remains a special kind of study, as it struggles to maintain a direct connection to the women's movement and to promote broad social change. It has been noted that if women's studies is the "academic arm" of the women's movement, that arm is in a sling. Women's studies was, and is, a central part of the academy at the same time that it continues to challenge and critique it. A tricky position to be in.

Despite its rocky beginnings, however, women's studies thrived in the 1970s. "A hundred and fifty new women's studies programs were founded between 1970 and 1975, and another hundred and fifty were founded between 1975 and 1980" (Boxer, 1988, 73). This period of growth coincided with numerous national policy-oriented and grant programs designed to ensure equity for women's education at every level, for example, the passage of Title IX—a federal law making sex discrimination in schools illegal—and the Women's Educational Equity Act (WEEA), which funded research, professional development, and a variety of resources to schools to bring attention to gender equity issues. In 1978, Congress included educational services in the Civil Rights Act designed to eliminate sex bias in school and society. This was followed in 1980 with a similar effort by the federal research agency known as the National Institute of Education (Sadker and Sadker, 1994, 36). The Fund for Improvement of Postsecondary Education (FIPSE) also funded a comprehensive evaluation of women's studies programs, which eventually became *The Courage to Question*, edited by former NWSA chair and longtime women's studies scholar, Caryn McTighe Musil.

Unfortunately, the 1980s, while a very fertile period for women's studies, were also subject to a conservative political backlash that women's studies had to weather. In addition, during the 1980s there was, and continues to be, a strong critique from women of color and other women who did not fit into the dominant definition of "woman." Another challenge in the 1980s involved the emergence of new feminist theories that were often at odds with each other. All this made the 1980s a particularly difficult yet exciting and prolific period for women's studies.

Women's Studies in the 1980s

Does a particular course foster a relational understanding of race, ethnicity, class, gender, and sexuality? Specifically, does it help students understand the significance of these categories? Does it identify how they operate individually? Does it identify how they operate in conjunction? (Silver, 1992, 162)

During the early 1980s, women's studies courses and programs grew at a rapid pace, with many colleges and universities offering interdisciplinary-based certificates and minors in the field, as well as the creation of many women's studies departments that offered full-fledged B.A.s. These programs, however, were severely attacked over the course of the decade in the face of the growing conservative nature of American politics under the Reagan administration. Many programs lost the bulk of their already modest funding and support, and many policies written to support equal rights both within and outside of academia were widely questioned. Those that survived—such as Title IX—have been under constant critique and should not be taken for granted. According to Weskott (2002):

> These attacks served to a link academic feminists to the objects demonized by right-wing Christians, as in Pat Robertson's now (in)famous fund raising-letter, "The feminist agenda is not about equal rights for women. It is about a socialist, anti-family political movement that encourages women to leave their husbands, kill their children, practice witchcraft, destroy capitalism and become lesbians." (297)

Pretty harsh words, indeed. Conservative critics both within and outside of the academy complained that women's studies "warped" the curriculum rather than transform it in a positive manner; that it was filled with "thought police"; and that it was totally focused on being "politically correct." By this it was meant that instead of being a legitimate source of knowledge about women in society, such programs silence discussion and debate.

Actually, during the 1980s, women's studies was undergoing an intensely self-reflective period as it grappled with the issue of how to identify the concept of "women," which had largely been defined as white, middle-class, heterosexual, Christian, educated women of privilege. *To be clear, questions about differences among and between women were already present in the late 1960s and 1970s*—both in women's studies and in related subjects that offered courses on women. For example, Toni Cade Bambara elected to teach a Black Women Writers class from her home after she was "prohibited" from teaching it in Spellman's English department. Students actually attended and took the course for no credit.

Yet questions about identity politics and women's diversity became more prominent, more troublesome, and more divisive in the 1980s. Women of color, black feminists in particular, noted that Women's Studies did not adequately include them or address their experiences and concerns. Many felt alienated to the point of wanting to create a new discipline or minor: Black Women's Studies. The term "woman of color" was also critiqued, as it suggests that women are either "the norm" meaning white, or the "other" meaning of color. And, of course, there are many differences among women of color.

Moreover, women's quest for equality with men is similarly complicated. As bell hooks has written: "Since men are not equals in a white supremacist, capitalist, patriarchal class structure which men do women want to be equal to?" (23). Though black women have been very vocal about their resistance to embrace feminism—as many believe that it is a largely racist movement that supports white women and renders black women as "other"—it is important to note, however, that it was not only black women and women of color who felt excluded from women's studies. What of woman who were "white" but still suffered from multiple axes of oppression—such as lesbian, bisexual and transgendered women, Jewish women, working-class women, and the like? Issues of ethnicity, nationality, language, religion, sexual preference, social class, and other differences among women were also points of contention as women's studies was largely the province of professors with little knowledge of or training in facilitating discussions about such difference.

The very definition of "woman" was now at stake. As will be explored in this chapter, and in the rest of the book, the issue of differences *among* women has been a critically important part of the development and evolution of Women's Studies. And it is important to note that, similar to the notion of "add women and stir" that suggests a quick-fix recipe for integrating women into the disciplines, dealing with the issue of difference within Women's Studies could/can be equally shallow. As Johnnetta Cole (2000) has self-consciously noted:

> My course Black Women in America, one of the earliest such sources, would today be evaluated as woefully lacking in sophistication. In those days, there was a tendency in all Black studies course to do what I used to call the Wheaties approach, that is, to identify the great figures in Black history and ask, in essence, why they were not on the front of a cereal box as were the white heroes. (330)

Just as feminism focuses on hierarchies of power between men and women, it must also look at the ways in which different groups of women are privileged in relation to each other. These questions were pervasive in the 1980s and are still highly relevant and controversial today despite a raised consciousness about cultural diversity throughout academia. It has even been suggested that women's studies needs nothing short of a "radical transformation" that, ironically, would reject the idea of gender as a primary category of analysis.

This may indeed be a project for the twenty-first century, as other scholars (e.g., see Judith Lorber in this book) have also suggested that we cannot think about gender as an essentialist category, that is, in isolation from differences *among* women. As bell hooks (1997) has noted: "When we cease to focus on the simplistic stance 'men are the enemy', we are compelled to examine systems of domination and our role in their maintenance and perpetuation" (25). I will come back to this quandary at the end of this chapter.

Despite widespread criticism and lack of adequate funding and support, in the 1980s, women's studies programs offering concentrations and minors were becoming more common and degree-granting departments were also growing at a rapid pace. As these women's studies became an actual department, and in many cases became more like a discipline, it was clear that it needed more organization. If women were going to get a B.A. in women's studies, they needed some sort of set of core courses—naturally an introduction to women's studies, but also at least one course in feminist theory.

Feminist theory came in many different flavors: liberal feminism; radical feminism; psychoanalytic feminism; cultural feminism, Marxist feminism/socialist feminism; standpoint epistemology; modern and postmodern feminism; and postcolonial feminism, to name a few. It is beyond the scope of this essay to describe each of these theories, yet the important point to be made is that, though full of interesting ideas, much of this theory was/is very difficult to understand, even for seasoned academics, much less for beginning students and feminist activists outside the academy. Moreover, many women's studies teachers were/are hesitant to teach feminist theory classes because they are trained in a particular discipline and do not feel comfortable teaching theory from other areas of expertise.

There has also been a stigma against "male" theory, as it has historically been used to oppress women. One teacher, for example, was very surprised when she walked into her first feminist theory class and not only found the students hostile, but "they were likely to accuse theory of being an active intimidation, manipulation, domination, hyper-rationalization, the evasion of feeling—and at the root of it all—an irredeemable maleness" (Cocks, 1985, 172).

The use of feminist theory was and remains a contentious issue in women's studies. The theoretical debates that emerged from both women's studies and the women's movement marked the tension between essentialism and constructivism, between the idea that women are "essentially" (e.g., born) different from men (beyond simply biology, in their ways of thinking and interacting with the world), and the idea that different cultures and circumstances foster and create such differences.

Two seminal women's studies theory texts that came out of the 1980s sparked a great deal of debate: *In a Different Voice: Psychological Theory and Women's Development* by Carol Gilligan (1982), and *Women's Ways of Knowing: The Development of Self, Voice, and Mind* by Belenky et al. (1986). Translated into nine languages and with 600,000 copies sold, *In a Different Voice* was a huge success and is still used in most feminist theory courses today. As the titles suggest, this books explore ways in which women and men are fundamentally different in their development, life experiences, and perspectives. Gilligan's research concluded that women act with an "ethic of care" and engage in "connected knowing," whereas men act with a focus on individual rights and justice.

In *Women's Ways of Knowing* the authors found a similar dynamic, noting that women were generally "subjective" or "connected" knowers who see truth as a result of one's personal experiences in the world, whereas men tend more toward "separate" knowing in which they make meaning strictly based on reason and disregard personal experience entirely. In both cases, these books underscored a notion of women's difference from men that was and continues to be contrasted with a notion of sameness and/or equality.

These books were both highly praised and widely criticized. While supporters noted that it was important to genuinely include women in research studies, and to be open to the idea that they are, in some respects, different from men, critics believed that the books simply underscored essentialist notions of men and women, without considering other important differences based on race, class, ethnicity, religion, sexuality, and so on. These texts, and others like them, remain important signifiers of the overall question of what unites and defines the category of women. Can we talk about "women" as if they are a cohesive category?

The common answer is no, we cannot. All women are not alike and the male/female opposition is not as tidy as one would think. In some circumstances, for example, black women may find they have more in common with black men than white women. Likewise, lesbians may feel more aligned with gay men than heterosexual women, and so forth. Perhaps most importantly, it needs to be acknowledged that women are not only oppressed in multiple ways, but that they can and have *even been oppressed by, and oppressors of other women.*

Feminist historian Joan Scott (1990) has written extensively on the subject in "Deconstructing Equality-Versus-Difference: The Uses of Poststructuralist Theory for Feminism." In this essay she makes the point that we must get beyond the binary opposition of equity (or sameness) versus difference. Her main point is that we must take each situation as a *specific context* rather than make blanket statements about men and women. She notes, for example that: "Political strategies...will rest on analyses of the utility of certain arguments in certain discursive contexts, without, however, invoking absolute qualities for women and men" (145). She is not denying that in many cases women and men respond and act differently, only that "its meanings are always relative to particular constructions in specified contexts" (145).

Also, to a certain extent, the debate around using feminist theory has mirrored the origins of women's studies in the academy. As noted throughout, women's studies was designed to work hand in hand with the women's movement and other movements for equity and social change. Yet, much of the theory coming out of women's studies programs was not only written in language inaccessible to most activists, it was difficult to see how the theory linked to actual practice. Students and faculty agonized about whether university work had any meaning in the "real world." Bowles (2002) has commented that: "From the beginnings of the field, feminist scholars wanted our academic work to have an impact on lives of women outside the academy.... in the 1980s, we agonized about whether university work had any meaning in the 'real' world" (458).

This fear was echoed by students too, as observed by Marjorie Jolles (2007) as she attempted to get her students to go beyond the common dichotomy of theory/practice. She notes that her students continually pondered whether feminist theory was accessible to those they believed to live in "a space they conjure as outside the classroom," for example, the "real world," and for whom they call "real women" (75). As Jolles notes, this comparison "obscures the many ways they share the world with 'real women'" (75). While Jolles is not denying class differences among college educated women and those who have not had access to this privilege, she does try to help her students to "consider that what they are doing in college is not demonstrating only a love of ideas but a kind of labor—indeed a form of practice...." (83). Jolles continues:

> The more they acknowledge their own participation in the university's project of the reproduction of class identity, the more they can develop a meta-stance towards the limits of the vocabularies and methods they are learning, the more they cultivate greater class consciousness. It is at this point that some are willing to admit they, themselves, are "real women" and that life in the university is very much a part of the "real world." (83)

While this realness must not be conflated with *sameness*, it is nonetheless important to challenge the idea that theory is only comprised of thought in an academic world and practice is merely action outside of the university. There are many overlaps here.

Many of these critiques of feminist theory in the 1980s continued into the 1990s and present day require serious consideration. Does feminist theory drive a gulf between learning and action, between the Ivy Tower and the community at large? Is most theory primarily jargon, and even more so elitist—comprised primarily of research studies and analysis of the lives of privileged educated white women? Is theory, as suggested above, mainly the province of males? Can we even define what it means to be a woman? If there is no such thing as a woman, what is feminism and women's studies without women? As described in the syllabus of one feminist theory course offered at San Diego State University called *Authority and the Politics of Representation in Feminism*:

> By the end of the course.... We will have explored the paradoxes in feminist theory's simultaneous insistence that women be represented in political discourse and practice, and its claim that "women" cannot be represented within the terms of dominant discourse because they cannot be defined.

The issue of whether women and men are fundamentally different, and if so, how to define and address those differences, continues. In the 1990s that debate branched out from the university to elementary and secondary schools. As the next section will explore, in the 1990s, women's studies began to expand its reach—both by focusing on girls and gender in K-12 settings—and by attempting to truly document and include the experiences of a wide diversity of women—and men!

Women's Studies in the 1990s

> Lowered self-esteem is a perfectly reasonable conclusion if one has
> been subtly instructed that what people like oneself have done in the
> world has not been important and is not worth studying. (AAUW
> Report, 1992, 67)

In the 1990s, women's studies underwent several important shifts,
one of which was that it began to focus much more readily on gender
equity and girls' experiences in the K-12 classroom. This was due
in large part to two publications: Myra and David Sadker's book,
Failing at Fairness: How America's Schools Cheat Girls (1994), and
the American Association of University Women's (AAUW's) report
How Schools Shortchange Girls (1992).

To be clear, there were some scattered, yet *very successful*, curricular and professional development programs aimed at pre-college
students and teachers in the 1970s and 1980s—most prominently
the Seeking Educational Equity and Diversity (SEED) Project (still
very active today), The New Jersey Project, the Dodge Seminars,
and The Piscataway Project. Florence Howe noted that as early as
1974 the Feminist Press imagined and designed women's studies
texts for the high school classroom that they hoped would rapidly
transform the curriculum, but learned that this was easier said than
done. In other words, it took a couple of decades for schools to
incorporate a lot of the exciting new material being developed in
women's studies, as this was not seen as part of the primary purpose and curriculum (e.g., reading, writing, and arithmetic), not to
mention the political dimensions involved. This is, in some ways,
still true today.

Yet, with the publication of the Sadkers' book and the AAUW
Report, the topic soon became cause for discussion and debate in
the popular media, on television shows such as *Oprah* and *Dateline*,
and widely read magazines like *Time* and *Newsweek* (as opposed
to academic journals). Throughout the decade many new programs
emerged in public and private schools, in all-girls' schools and coed
schools, in urban, suburban, and rural schools. Similar to the issues
raised in the university, some of the issues of concern in K-12 education
include: whether girls and boys got similar attention from teachers;
whether the curriculum included the experiences and achievements of
women—both famous and not famous; whether girls were being discouraged from pursuing typically male fields like math and science;

whether girls were subject to greater levels of sexual harassment; whether girls suffered from greater loss of self-esteem as they often felt silenced in the classroom, and many others.

In their book, the Sadkers tell riveting stories such as the use of textbooks that in over a thousand pages devote only a single line to women's suffrage. They also note that when a group of students were asked, "Suppose you woke up tomorrow and found you were a member of the other sex. How would your life be different?" While girls were at least open to the idea, the Sadkers reported that "For boys the thought of being female is appalling, disgusting, and humiliating; it is completely unacceptable." A number of boys even said they would kill themselves (82). When fourth, fifth, and six graders were asked to write down the names of as many famous women and men as they could in five minutes, "[o]n average, students generated eleven male names but only three women's and they tended to be less from the pages of history than from popular culture—such as Princess Di and Aunt Jemima" (71). The Sadkers also noted that "Tolerated under the assumption that 'boys will be boys' and hormone levels are high in high school, sexual harassment is a way of life in America's schools" (9).

The AAUW report echoed many of these same findings. Among the report's many recommendations was that: "Support and released time must be provided by school districts for teacher-initiated research on curricula and classroom variables that affect student learning; Gender equity should be a focus of this research and a criterion for awarding funds" (85), and, further that: "Teachers, administrators and counselors should be evaluated on the degree to which they promote and encourage gender-equitable and multicultural education" (67).

In a three-year professional development program for urban teachers, parents, and administrators around gender awareness titled, Gender Awareness through Education (GATE), teachers raised questions such as: "Do I point out how girls differ from boys, or just encourage them without making a big deal of my perceived differences?"; "When we say it's harder to work with girls aren't we saying something about ourselves?"; "How can we get to know students personally without invading their privacy?"; and "How does being 'powerless' cut across all lines?" (Ginsberg, Shapiro, and Brown, 2004, 46). These are not easy questions to answer, especially as these teachers taught in urban public schools with large percentages of minority and disadvantaged children for whom gender was only one area in which they were "oppressed."

AAUW went on to publish a whole series of reports about gender equity in public education, including *Girls' in the Middle* written by Research for Action (1996)—which focused specifically on girls in middle school settings. This report concluded that schools should take an Educational Equity Inventory, which poses such critical questions as: Does your school have "a sexual harassment policy"; "data that show student achievement by gender"; "forums for dialogue across the various constituencies of the school"; "support in the school and the district for adults who mentor girls"; and many more. Most schools did not have such policies in place. It was also found that middle school was a particularly difficult time for students, developmentally in general, and certainly adding questions about gender did not easily resonate with students, many of whom were perpetually silent in class, and for whom feminist pedagogy was a huge change in structure from the more tried and trusted "memorize and be tested on" curricula.

In one GATE middle school, a science teacher asked her students to list seven male scientists and seven female scientists and to write a report on one of each. The students couldn't come up with seven female scientists. When asked why they thought this was, the boys answered: "I think we probably didn't find more women because women are lazy"; "Women aren't fit to be scientists, especially not a rocket scientist! It's too dirty for women and they're not strong enough anyway"; and "Girls have to spend so much time looking pretty that they can't spend the long hours it takes to be a scientist." (Ginsberg, Shapiro, and Brown, 2004, 111–112).

Trying to introduce women's studies into the high school curriculum was equally illuminating in terms of the resistance of boys to embrace feminist ideas. One teacher noted, for example, that boys were so afraid of seeming feminine that they made crude comments such as some women were even too ugly to be raped.

Despite boys' resistance to embracing women's studies curricula, raising issues about girls' education nonetheless introduced a dialogue about boys' education. Newly stressed attention to boys was evident in such as examples as the fact that The Ms. Foundation changed its highly praised "Take Your Daughters to Work Day" to "Take Your Daughters and Sons to Work Day," recognizing that men will not start treating women differently unless they see role models of women in leadership positions, particularly in traditionally male fields. Likewise, the follow-up to the AAUW report on *Shortchanging Girls* was titled *Gender Gaps: Where Schools Still Fail our Children*—conspicuously leaving "girls" out of the title.

Especially in poor, urban schools with large minority populations, attention to underprivileged boys is certainly a worthy pursuit. But, of course, one does not have to occur at the cost of the other. While research suggests that there is reason for concern that boys—particularly marginalized boys—are getting a quality, equitable education, it has become a sort of contest between boys and girls as to who is more "at risk," rather than looking at the fact that both groups can benefit from new teaching strategies and educational environments. Moreover, as I will explore in the next section on the twenty-first century, there lay a danger that females would, once again, be marginalized. It has, in fact, become more and more commonly suggested—even by some feminists themselves—that we've "solved" the problem for girls and women. While great progress has been made, we have a long way to go on this front.

Meanwhile, in the academy another prominent issue in the 1990s concerned the issue of how to include sexuality studies in women's/ gender studies. Although it had always been a part of women's studies and gender studies, it was mostly centered on two categories: lesbians and gays. It did not include to any great extent bisexuals, transsexuals, or transgendered people. In the 1990s came the development of what was to become known as "queer theory." As Nancy Rabinowitz (2002) has noted: "[B]efore 1991 queer theory can hardly be said to have existed, but by 1996 it was a common parlance" (2002, 178). According to Rabinowitz, one of the reasons the title was not used widely in the past was that students did not want it on their transcripts. And yet, as she also notes, queer is more inclusive than the label gay/lesbian, and that "On the simplest level, the term provides an umbrella so that we don't have to keep adding identities or resort to LGBT"W" (for whatever....)." Rabinowitz concludes that "A queer perspective would.... problematize the subject of sexuality in general, and would reveal structures of heteronormativity" (183).

Rabinowitz also raised the question of whether she, as a heterosexual, should be able to teach courses on queer theory. This question has been posed more broadly, as to whether the women's studies mantra that "the personal is political" means that if you have not experienced something personally, you are not fit to speak about it, or challenge others views on it. Sometimes referred to as "standpoint epistemology," women's studies professors are often challenged by expectation that they will only teach courses on subjects with which they personally identify. For example, only a black woman can teach a course of black women's studies. According to Amie Macdonald (2002), who has written about this issue extensively in "Feminist Pedagogy and the Appeal to Epistemic Privilege":

> Because I precast the epistemic framework of the course through self-evident experiential knowing, I effectively undercut my ability to generate critical standpoints on the experience-based analysis of my students who occupy subject positions that differ from my own. (114)

She also rightly points out that all students of a certain identity do not experience that identity in the exact same manner, so, in effect, nobody can speak for an entire group. According to Macdonald:

> When one member of an oppressed social group claims epistemic privilege, others often presume that all similarly situated people share the same viewpoint. Ironically, this often has the effect of reinscribing racist, sexist, homophobic, and classist ideologies. (119)

She suggests that students form *communities of meaning* that are not so much organized around "identity markers" such as race, class, gender, sexuality, religion, or nation, but instead by "shared experiences and ways of understanding and being in the world" (120). In this framework, "students and faculty would decide for themselves which community or communities they wanted to form or ally themselves within any given context" (120).

Thus, the personal is still very much political, but we (feminists/women's studies students and teachers) are not bound by the personal.

Also in the 1990s, the age-old question of whether or not Women's Studies should become a "discipline" in its own right despite its inherently interdisciplinary nature persisted. Again, this question was present from the early founding of women's studies, but in the 1990s it became particularly controversial. Inherent to the question of whether Women's Studies should be its own discipline are other questions, such as:

- Does making Women's Studies its own discipline risk ghettoizing it or does it further validate it and secure its future?
- Does Women's Studies, like African American studies, urban studies, religious studies, and other interdisciplinary fields, set out to critique the very notion of disciplines and compartmentalization of knowledge?
- Is Women's Studies the only place where overtly feminist classes are offered?

- Have we given up on trying to "transform" the university?
- If Women's Studies genuinely becomes an integral part of other disciplines, will there be no reason for it to continue on its own?

Also, as Jean Fox O'Barr (1994) has wisely asked: "What does Women's Studies have to offer other disciplines in the academy? At what price to itself?" (282). In other words, if women's studies became a legitimate department within the university, would it still be able to retain its close ties to the feminist movement?

So, should women's studies be its own discipline? The jury is still out. It is likely that the question of whether women's studies should be a discipline will likely plague the field for decades to come, however, also in the 1990s, colleges and universities began to offer more graduate program options in women's studies. These included certificates, masters, and doctoral programs. This development further complicated the debate around whether or not women's studies was or should be a "discipline" in its own right.

Women's Studies in the Twenty-First Century

> Should women's studies Ph.D. programs be concerned with transforming institutional politics and values, both internal and external, as well as the conventions of Ph.D. education itself? (Arenal, 2002, 190)

In the 1990s and the twenty-first century, women's studies remains a popular option, and continues to grow as a field. Some women's studies classes have a waiting list as high as 200 students! The emergence of graduate options and online learning (both of which will be discussed herein) are central factors in this growth. However, the field is still fraught with questions raised throughout this essay, questions that will likely continue deep into the twenty-first century, such as, how to define "woman" and what is the primary agenda of structure of women's studies—if indeed there is a common answer to this question.

One of these questions that has not yet been raised in detail is the role of males in women's studies. This is a role that is growing, as the field questions whether it even wants to remain being called *women's studies*. In the 1990s and the twenty-first century, women's studies programs have played around with alternate words such as "Female

Studies," "Womanist Studies," "Gender Studies," and "Sexuality Studies." Many programs have added gender studies to the name, or dropped Women's Studies altogether in favor of gender studies alone. In addition to the idea that gender is socially constructed, the change to *gender studies* suggests that we need to be paying attention to the relationships between men and women rather than focusing predominantly on women's experiences and knowledge itself. It has also been suggested that gender studies is a more appropriate title because it is more inclusive of gay, lesbian, and transgendered individuals. On the other hand, taking the *women* out of women's studies poses some serious risks, the most obvious being that gender studies is not *necessarily* feminist in nature, nor is it necessarily linked to the women's movement, as there is no "gender movement."

Recently, Haverford College and its sister school Bryn Mawr College in Pennsylvania polled their students, faculty, and administrative staff to vote for a name for their Women's Studies program, and to provide comments on why they voted as they did. Their choices were: 1) Feminist & Gender Studies, 2) Feminist, Gender and Sexuality Studies, 3) Gender and Sexuality Studies, and 4) Gender Studies. The name "Gender Studies" won by a significant margin. Although most took the project of naming very seriously, a few people wrote things such as "Just name it—nobody cares." And yet, there are reasons to care.

As will be raised many times in this book, controversies as to what to call Women's Studies illuminate the ways in which the field has had to continually reassess its purpose, audience, and subjects. As research will show, while still tied to its activist roots, the switch to gender studies has been controversial, with many students suggesting that Women's Studies and feminism—both focused overtly on women—is a relic of the past. Some key comments along this line included: "I think the term 'feminist' tends to scare off people," and "Feminism is an obsolete term.... women today are no longer oppressed."

At the same time, questions must thus be raised: Does changing the name from Feminist or Women's Studies to Gender Studies depoliticize its primary goals and work?; Does taking out the word women make the department less threatening, more inclusive, and more mainstream, and/or does it actually hold men accountable as gendered beings? Can there be a women's studies without "women"? Is women's studies the proper place to study sexuality studies? Does adding sexuality studies to the title make Women's Studies less essentialist (e.g., men versus women), making way for transgendered individuals, and building an inviting space for lesbian and gay students at the same time? Perhaps most importantly, what are the implications of including sexuality under the rubrics of Women's Studies, but not (explicitly) race or class?

These are no easy questions to grapple with, for sure. On a Women's Studies list-serve the question was raised: "Does anyone know any good pieces on why it important to call the discipline Women's Studies rather than Gender Studies, Women and Gender Studies, etc.?" The response was widespread and I can only touch briefly on it here. One person responded: "I fear a depoliticizing of feminist scholarship if we shift too far away from the study of woman." Another wrote: "It's possible that by insisting on Women's Studies exclusively we collude in perpetuating the notion that men don't have a gender, that they are a universal 'norm' studies in the 'regular curriculum' while women are studies in this special area of the curriculum." Yet another wrote that: "We need to destabilize gender at the same time we insist that historically and politically a category or class of individuals called women have systematically been oppressed." This is a worthy goal, but nonetheless a complex and most difficult one to accomplish.

The question remains, can we have a women's studies without overtly focusing on feminism? Zimmerman (2005) has written:

> There is no women's studies without feminism.... Feminism is what turns the study of women into women's studies. Gender studies might or might not be feminist. Gay and lesbian, or sexuality studies might or might not be feminist. Ethnic studies or cultural studies might or might not be feminist. But women's studies must be feminist or it is not women's studies. (37)

The question of what to call women's studies has continued into the twenty-first century and is not likely to go away any time soon, as students and teacher grapple with what it means to engage in political discourse in the university.

These questions are even more complicated by the fact that graduate courses and graduate programs in women's studies are now growing institutionally across the country. Such programs include certificates, masters, and doctoral degrees, though many of them are designed to be an addendum to graduate training in other disciplines. They are also titled in many different ways, including: gender studies, feminist studies, women's studies, and Gender, Women and Sexuality Studies. Presently, there are over 60 graduate certificate programs in women's studies. American-based masters programs in women's studies went from 6 in 1994 to over 34 in 2007. Between 1997 and 1998 the number of women's studies Ph.D. programs increased by 50 percent (Davis, 2007, 272). In the twenty-first century, there are still only a handful of opportunities to get a Ph.D., but this handful is growing. And they are surprisingly very popular. At the University of Maryland, every year there are about 100 applicants to the Ph.D. program, of which only five to seven are commonly accepted.

While it is a very exciting development in the field, creating the potential for students who feel passionately about women's studies to go to the next level and for a new generation of women's studies scholars, truly trained to do interdisciplinary work, to be born, is not without its problems. May (2005) has noted, for example, that the Ph.D. in women's studies "connects to the field's ongoing debates around excess." By this she means "the alleged impossibility of 'doing everything,' of being accountable to multiple methods, constituencies, identities, geographies, and activisms" (185). Davis (2007) has similarly noted that:

> In asking whether or not women's studies Ph.D. programs should exist, a central issue addressed by the 1998–99 debates was the disciplinary capacity of women's studies to produce knowledges demonstrative of the expertise and specialization required at the level of doctoral training. (273)

As women's studies continues to be an interdisciplinary enterprise, a Ph.D. in women's studies has the potential to radically change the way we think about the degree itself that has traditionally been a way to delve much more deeply into a *specific disciplinary content*. Perhaps for this reason, those institutions that do offer doctoral degrees solely in women's studies, most often offer students a choice of two or three concentrations—such as: Language, Literature, and the Arts; History, Psychology, and Society; Sexualities, Desires, and Identities; Cultural Representations and Media Practices; Women in an International Context; Women, Leadership, and Public Policy; Feminist Theory; Comparative Gender Roles; Women of Color; Transnational Feminisms—rather than leaving the field wide open. In addition, many women's studies, Ph.D. programs require a service-learning component that links academic learning to activism. One such program is at the Union Institute—one of the oldest institutions offering graduate programs in women's studies—which requires an internship of 500 hours or the equivalent of three months of full-time study.

There is also the pervasive question about what one *does* with a Ph.D. in women's studies, as May (2005) and others have observed, that while many advertisements ask for *interdisciplinary* teaching experience, they still favor hiring candidates with disciplinary scholarship or degrees. On a women's studies list-serve the question is raised whether a particular student should go on to get a graduate degree in women's studies in spite of the fact that her advisor has cautioned her that she will have trouble getting a job. While there is not total agreement on the subject, other posts to the list-serve confirm that most academic departments—even women's studies departments!—are hesitant to hire candidates with women's studies degrees (e.g., as opposed to specializing in the study of women within a particular discipline.) One person wrote, for example:

> ...I now realize, having been on the job market, that having a Ph.D. in WS is not all that I thought it would be. Traditional departments (my other area is sociology/criminology) often demand a Ph.D. in their discipline. They often view the WS degree with some suspicion.... I have had interest from interdisciplinary sociology programs, but these are few and far between.

This opinion, however, is not shared by everyone. Some argue, for example, that having an interdisciplinary background is helpful in a job market that is highly competitive and in which professors are able to teach a variety of courses. Such programs are growing, and hold the potential to create a new model of women's studies, as a new generation is educated outside of the traditional "disciplinary" model. Moreover, these programs have produced many interesting and diverse masters theses and dissertations, a very small sampling of which includes:

- "Passing the Talking Stick: America Indian Women's Voices Heard" (Dawn Cicero, Fall 1999)
- "'Our Politics Was Black Women': Black Feminist Organizations, 1968–1980" (Kimberly Springer, Summer 1999)
- "Beyond Race and Gender? The New Managing Diversity for Women: A Dual Approach" (Jolna Karon, Spring 2003)

Finally, another development in the twenty-first century is the use of computers to facilitate women's studies classes, discussions, debates, and even online degree programs! While women's studies list-serves existed in the 1990s, they have grown rapidly and have expanded to entire courses being taught online. Notes Whitehouse (2002):

> All traditions of teaching are now challenged by new ways of thinking, knowing and learning through the technology of innovation that permeates our school system from Kindergarten to higher education. (209)

The ability to offer women's studies courses and discussions online makes the field considerably more accessible for working parents and others who find it difficult to attend classes in person. It also alleviates questions of "time" that constrain the actual classroom, as in the virtual classroom—while there still may be real-time discussions and deadlines—there is the potential for greater flexibility overall.

Online learning has also been effective in creating communities to supplement university courses and give voice to those who may be traditionally silent. Although girls and women have traditionally been discouraged from pursuing new technologies, the time has come for this to change. Whitehouse (2002) define the characteristics of an online community as encouraging active interaction and collaborative learning, dialogue, shared resources, encouragement, and support (214).

It has been noted that such communities are student centered, with faculty acting as facilitators in most cases. In one women's studies online course, the professor created a class Web site that had a page called the *Who's Who page* in which students have a template for talking about their personal experiences, politics, and interests. What is important about this is that the page can be edited throughout the semester as "many students make changes in their pages to reflect progress in their thinking" (Whitehouse, 2002, 218).

Perhaps more importantly, however, in online courses students who are apt to dominate class discussions cannot as easily dominate asynchronous discussions, leaving more room for normally shy students to contribute. One student wrote of the experience of taking a women's studies class online:

> I highly recommend the online courses! Everyone has a chance to participate—in a non-threatening fashion. Sometimes people feel inhibited in a classroom, but this format encourages articulation and exploration of thoughts and ideas. You get to connect with people all over the country (world) in ways you can't in a [face to face] class. That

allows different experiences and regional variations to surface, which only adds to the curriculum.

Another noted:

> Even though the class "meets" at the discussion board and not face to face, each person's personality and individuality shows through.... Each person responds to the readings in a unique way, adding new elements to my learning process. In absence of a lecture, where only one point of view is presented, we each are treated to the life experiences and knowledge of each person in the course. You can't buy that in a book. "Each one teach one" is an adequate description of WS online graduate courses.

The interest of pursuing women's studies online has spread to emerging opportunities to get online degrees, such as a graduate certificate offered at Southern Connecticut State University (SCSU), one of the longest running women's studies departments in the nation. Students meet over the summer for a required two-week residency on SCSU campus, during which time, as noted on their Web site, "students take a course in Women's Studies Foundations and build relationships with faculty and other Women's Studies students." The remaining credits are earned during the academic year. It is stressed, however, that the two-week residency is an important part of the program as "The on-ground component of this new certificate program provides what is often missing in a distance-learning environment and helps the cohort forge a sense of community, an important element in [the] women's studies classroom" (http://www.southernct.edu/womensstudies/).

This finding was actually unearthed in the 1990s, when Ellen Rose taught her first class in Cyberspace. Groups of students from the University of Nevada, Reno (UNR) and the University of Nevada, Las Vegas (UNLA) enrolled in a classroom that met via teleconferencing and email. According to Rose (1998), most of the class students were very unhappy, as they came to the class with very different exposures to women's studies, to the technology involved, and to each other and the teacher, who was based in Las Vegas. She described the email process as a "war" (you say/I say) and emphasized students' discomfort with talking to each other via a video camera. Indeed, for the UNR group (which had practically no prior exposure to women's studies, as opposed to the UNLA group that was primarily comprised of women's studies majors), the television screen was positioned in such a manner as it was literally, as well as metaphorically, looking down on the students.

Yet even in the 1990s, Rose said she would do it again, only next time she would make sure that the students had an opportunity to meet face-to-face at the beginning of the semester—regardless of the costs involved. She also noted that if she were to use email again she would do more "to make it an integral part of the course" (Rose, 1998, 128) rather than use it solely as an opportunity for students to dispute differences.

As technology continues to expand, it is likely that options for online courses and degrees as well as interdisciplinary discussions and debates will blossom. The caveats noted above will be important, as online learning is a difficult task for those trying to teach using feminist pedagogy. It will remain essential that the class feels a connection, and an accompanying responsibility for each other's lives and learning, regardless of their differences in distance and demographics.

In conclusion, one of the most pressing issues facing women's studies today continues to be the issue of essentialism versus constructivism—that is, whether women are inherently different than men in some respects or whether these differences are socially and politically created. I mention this here again, because in the twenty-first century there have been a growing number of scholars who believe that woman and, more specifically, *gender* itself are not useful categories of analysis. Though this has the potential to fraction its relationship with the women's liberation movement, Zimmerman (2002) has noted:

> We will need to analyze the relationship between academics and activism, between Women's Studies and women's movements, in light of the social realities of the early twenty-first century, not our nostalgic memories of the 1960's. (187)

Lorber in particular has written about this in her newest book, *Breaking the Bowls: Degendering and Feminist Change.* As noted throughout the book, this means going "beyond gender" by acknowledging the many differences among women based on race, class, ethnicity, sexuality, as well as age, parental and relational status, physical ability, education, and religion. Notes Lorber (2005):

> This multiple perspective fragments gender, and breaks the hold of binary categorization. I think that for feminists in modern Western civilizations, going beyond gender is a needed step towards gender equality.... (155)

These questions will continue into the twenty-first century. Yet this is not a negative thing. As Du Bois (1983) has rightly noted:

> There is no question that feminist scientists and scholars will continue to be charged with bias, advocacy, subjectivity, ideologizing, and so on. We can expect this: we can even welcome it. If our work is not in some way threatening to the established order, we're on the wrong track. (112)

References

Aiken, S., K. Anderson, M. Dinnerstein, J. Lensink, and P. MacCorquadale. (1988). *Changing our minds: Feminist transformations of knowledge.* Albany: State University of New York Press.

Alcoff, L. (1988). "Cultural feminism versus post-structuralism." In E. Minnich, J. O'Barr, and R. Rosenfeld (Eds.), *Reconstructing the academy: Women's education and women's studies.* Chicago: The University of Chicago Press, 257–288.

American Association of University Women (AAUW). (1992). *How schools shortchange girls.* Washington, DC: Joint publication of the American Association of University Women and the National Educational Association.

———. (1999). *Gender gaps: Where schools still fail our children.* New York: Marlowe and Company.

Arenal E. (2002). "Implications and articulations: The Ph.D. in Women's Studies." *Women's Studies Quarterly*, 30(3 and 4): 179–192.

Belenky, M.F., B.M. Clinchy, N.R. Goldberger, and J.M. Tarule. (1986). *Women's ways of knowing: The development of self, voice, and mind.* New York: Basic Books.

Bowles G. (2000). "From the bottom up: The students' initiative." In F. Howe (Ed.), *The politics of Women's Studies.* New York: The Feminist Press, 142–154.

———. (2002). "Continuity and change in Women's Studies." In R. Weigman (Ed.), *Women's Studies on its own.* Durham and London: Duke University Press, 457–464.

Boxer, M. (1988). "For and about women: The theory and practice of Women's Studies in the United States." In, E. Minnich, J.F. O'Barr, and R. Rosenfeld (Eds.), *Reconstructing the academy: Women's education and Women's Studies.* Chicago: University of Chicago Press, 69–103.

———. (2001). *When women ask the questions: Creating Women's Studies in America.* Washington, DC: Johns Hopkins University Press.

Brazin, N.T. (2000). "The gender revolution." In F. Howe (Ed.), *The Politics of Women's Studies.* New York: The Feminist Press, 57–68.

Cocks, J. (1985). "Suspicious pleasures: On teaching feminist theory." In M. Cully and C. Portuges (Eds.), *Gendered subjects: The dynamics of feminist teaching.* Boston: Routledge & Kegan Paul, 171–194.

Cole, J. (2000). "The long road through gendered questions." In F. Howe (Ed.), *The politics of Women's Studies*. New York: The Feminist Press, 327–333.

Davis, D.R. (2007). "A new wave, shifting ground: Women's Studies PhDs and the feminist academy from the perspective of 1998." In H. Aikau, K. Erickson, and J. Pierce (Eds.), *Feminist waves/feminist generations: Life stories from the academy*. Minneapolis: University of Minnesota Press, 270–289.

Dinnerstein, Myra. (2000). "A political education." In F. Howe (Ed.), *The politics of Women's Studies*. New York: The Feminist Press, 291–305.

Du Bois, B. (1983). "Passionate scholarship: Notes on values, knowing and method in feminist social science." In G. Bowles and R.D. Klein (Eds.), *Theories of Women's Studies*. London: Routledge and Kegan Paul, 105–116.

Farley, T.P. (2000). "Changing signs." In F. Howe (Ed.), *The politics of Women's Studies*. New York: The Feminist Press, 264–275.

Gilligan, C. (1982). *In a different voice: Psychological theory and women's development*. Cambridge, MA: Harvard University Press.

Ginsberg, A., J. Shapiro, and S. Brown. (2004). *Gender in urban education: Strategies for student achievement*. Portsmouth, NH: Heinemann.

Guy-Sheftell, B. (2000). "Our mothers of Women's Studies." In F. Howe (Ed.), *The politics of Women's Studies*. New York: The Feminist Press, 216–226.

hooks, b. (1997). "Feminism: A movement to end sexist oppression." In S. Kemp and J. Squires (Eds.), *Feminisms*. Oxford: Oxford University Press, 22–27.

Howe, F. (1997). "The first ten years are the easiest." *Women's Studies Quarterly*, 25(1 and 2): 23–33.

Jolles, M. (2007). " 'Real women' in Women's Studies: A reflective look at the theory/practice dilemma." *Feminist Teacher*, 18(2): 74–85.

Lorber, J. (2005). *Breaking the bowls: Degendering and feminist change*. New York: W.W. Norton and Company.

Macdonald, A. (2002). "Feminist pedagogy and the appeal to epistemic privilege." In A. Macdonald, and S. Casal (Eds.), *Twenty first century classrooms: Pedagogies of identity and difference*. New York: Palgrave, 111–133.

May, V. (2005). "Disciplining feminist futures?: 'Undisciplined' reflections about the Women's Studies Ph.D." In E. Kennedy and A. Beins (Eds.), *Women's Studies for the future: Foundations, interrogations, politics*. New Brunswick: Rutgers University Press, 185–206.

Musil, C. (1992). *The courage to question*. Washington, DC: Association of American Colleges and National Women's Studies Association.

O'Barr, J.F. (1994). *Feminism in action: Building institutions & community through Women's Studies*. Chapel Hill: University of North Carolina Press.

Rabinowitz, N. (2002). "Queer theory and feminist pedagogy." In A. Macdonald and S. Casal (Eds.), *Twenty first century classrooms: Pedagogies of identity and difference*. New York: Palgrave, (175–200).

Research for action. 1996. *Girls in the middle: Working to succeed in school*. Washington, DC: American Association of University Women Foundation.

Robinson, Jean. (2002). "From politics to professionalism: Cultural change in women's studies." In R. Weigman (Ed.), *Women's Studies on its own*. Durham and London: Duke University Press, 202–210.

Rose, E.C. (1998). "This class meets in cyberspace: Women's Studies via distance education." In G. Cohee, E. Daumer, T. Kemp, P. Krebs, S. Lafky, and S. Runzo (Eds.), *The feminist teacher anthology: Pedagogies and classroom strategies*. New York: Teachers College Press, 114–132.

Sadker, M. and Sadker D. (1994). *Failing at fairness: How America's schools cheat girls*. New York: A Touchstone Book.

Scott, J. (1990). "Deconstructing equality-versus-difference: Or, the uses of poststructuralist theory for feminism." In M. Hirsch and E.F. Keller (Eds.), *Conflicts in feminism*. New York and London: Routledge, 134–148.

Silver, L. (1992). "Oberlin college: Self-empowerment and difference." In K. Musil (Ed.), *The courage to question*. Association of American Colleges and National Women's Studies Association, 157–177.

Weigman, R. (2002). *Women's Studies on its own*. Durham and London: Duke University.

Weskott, M. (2002). "Institutional success and political vulnerability: A lesson in the importance of allies." In R. Weigman (Ed.), *Women's Studies on its own*. Durham and London: Duke University, 293–311.

Whitehouse, P. (2002). "Women's Studies online: An oxymoron?" *Women's Studies Quarterly*, 30(3 and 4): 209–225.

Zimmerman, B. (2005). "Beyond dualisms: Some thoughts about the future of Women's Studies." In E.L. Kennedy and A. Beins (Eds.), *Women's Studies for the future*. New Brunswick, NJ: Rutgers University Press, 31–39.

Part Two

Reflections

1

A Personal and Epistemological Journey toward Women's Studies

Margaret Smith Crocco

Introduction

Alice Ginsberg's invitation to write this chapter presented authors with a number of questions to which they were asked to respond. I have chosen to take an autobiographical approach to her request since personal experiences, rather than formal academic training, brought me into women's studies, changed my thinking about "how we know what we know," and shaped my life. My "participant observer" status as an American woman along with 30 years of reading and teaching about women's status worldwide continues to sustain my interest in a burgeoning field that has, since its inception, merged a political with an intellectual agenda. I hope the reader will indulge me in this approach, a ploy entirely in keeping with two of the fundamental insights of the U.S. feminist movement—that the personal is the political and that one's positionality shapes one's intellectual work.

The Early Years

In 1976, fresh out of the Ph.D. program in American Civilization at the University of Pennsylvania, I began my academic career teaching at the University of Maryland. The most popular undergraduate offering on campus that year was a popular culture course on soap opera, which attracted over 300 students. My own course, Death and Dying in American History, pulled in a respectable 100 students, although my two other courses—one on life histories and another on age and sex roles in American culture—were, happily for this novice, a bit smaller. Although I spent only a little more than a year teaching in the American Studies program at Maryland, it was an experience filled with new perspectives on what it meant to be a woman in academic life and to juggle marriage, family, and career in a society unsupportive at best and hostile at worst to working mothers.

The product of a large family with five sisters and two brothers, I was fortunate to have parents with an egalitarian, open-minded approach to childrearing. I was a graduate of Oak Knoll School of the Holy Child, a Catholic, all-female high school run by nuns who took women's intellectual capacities seriously, and Georgetown University, a Jesuit undergraduate institution that had only recently allowed women to matriculate into the College of Arts and Sciences from its professional schools. Thus, I came to my convictions about what women could be and do, largely through socialization and reflection on my own and others' experiences. I did not have the benefit of women's studies courses nor on-the-ground training in a radical women's group. When I began graduate school in 1972, I had read neither Simone De Beauvoir nor Betty Friedan.

My college and graduate school education had taught rather different lessons concerning women's place in academic life. Although virtually all of my high school teachers were women, at college I had no female professors. I should point out that I had made a deliberate choice not to attend a women's college, but it did become clear during my undergraduate years that there were certain trade-offs with this choice. It was not that I can remember any overt discrimination at Georgetown. Indeed, I was surrounded by an extraordinary group of women who were—uncharacteristically for that time—pursuing career preparation in the foreign service, law, and medicine. Although I was conscious of our status as pioneers in the College of Arts and Sciences at Georgetown, my college years—1968 to 1972—were more focused on opposition to the presence of ROTC (Reserved Officer Training Corps) on campus and the war in Vietnam than on women's issues.

When I began at Georgetown, women were required to wear skirts to classes and encountered "in loco parentis" regulations governing dorm and college life. Despite the paternalism, I valued my education there, having met a number of bright students and several professors who encouraged me in my pursuit of graduate education.

When I arrived at Penn, I was happy to find several female instructors there. As I was finishing my doctorate four years later, the first appointment of a woman to a tenure-track position in the American Civilization Department was made—Drew Faust, who became the first female president of Harvard University in 2007. Clearly, hide-bound academic institutions change very slowly, especially those with long traditions of calculating merit in narrow and self-preserving ways.

The American Civilization Department was most helpful in opening up new ways of thinking about merit, excellence, epistemology, and appropriate questions and topics for academic study. With its unifying approach based on the culture concept—whether directed toward material objects, media, religious and social communities, or political institutions and practices—our program of study broke down the distinctions between what have come to be known as "high culture" and "low culture." We were interested in analyzing both, considering them cultural products. We discussed Thomas Kuhn's (1962) *Structure of Scientific Revolutions* and applied the notion of paradigms and paradigm shift quite consciously to our own epistemological approach to American "civilization," nomenclature reflecting the fact that, at least titularly, we had not broken free of older ways of thinking.

My focus on history and anthropology provided a conceptual framework for critiquing what might be meant by concepts such as civilization, high culture, and low culture and the implicit hierarchies such terminology implied. We were also schooled in using cross-cultural and historical comparison to register the notion that social and cultural arrangements were not universal. We read Alfred Kroeber, Anthony F.C. Wallace, James Spradley, Clyde Kluckholn, and Margaret Mead. Intellectually smitten with anthropology, I wrote my dissertation under an anthropologist, Dr. John Caughey. My study, "Adolescence as Artifact: An Ethnography of a Psychiatric Clinic," allowed me to put knowledge and skills to work on a critique of age and sex role arrangements of modern American society.

During my year teaching at Maryland, several important events occurred that catalyzed my move toward women's studies. I was invited to attend a meeting of professors interested in organizing a women's studies program on campus. As I recall, we met in the basement of a local restaurant, perhaps two dozen women from across the arts and sciences. We were all interested in studying women as a legitimate academic enterprise; we planned on developing a support group for female academics college-wide and considering what collective action was needed on campus to make this happen. We were the vanguard of the women's studies movement at Maryland

The mission statement of the contemporary Department of Women's Studies there captures the flavor of what we were about in the 1970s:

> We see our department as a force for change in the world, change which strives to achieve intellectual freedom, social justice, and equality for all people. We do this by providing an outstanding education in women's studies through excellent teaching, engaged mentoring, path breaking research and scholarship, and dedicated community service. (www.womenstudies.umd.edu/about/missionstatement.shtml)

The mission statement goes on to discuss interrogating human differences "such as those of gender, race, class, sexuality, nation, ability, ethnicity, and religion." This formulation reflects 30 years of women's studies scholarship that has moved the field from its narrow origins in white, middle-class American "womanhood" to an orientation inflected by difference in multiple, often challenging, ways.

Later that year, provocative visual artist Judy Chicago exhibited her work, "The Dinner Party," on campus. I remember being stimulated and even moved by the scene: a good sized room, a large, triangular shaped table in it with about a dozen place settings related to women in history. Each setting included a placemat or piece of tablecloth with a woman's name on it; each plate featured a butterfly or flower-like object connoting women's reproductive capacities.

The work had emerged from the collaboration of a group of women between the years 1974 to 1979; the completed work was first exhibited formally in 1979. What we saw at Maryland was a preview of the completed project, a true work-in-progress that was stunning in its representation of women's forgotten past. Judy Chicago later described the purpose of this work as a means "to end the ongoing cycle of omission in which women were written out of the historical record." Few of these names, except for the artist Georgia O'Keefe, were familiar to me, but the exhibit helped shape the largely inchoate set of life experiences I had had into something that might plausibly be called a feminist epistemological stance: Who were these women? And why hadn't I heard about them before? How could my education have overlooked their accomplishments?

Betty Friedan (1963) called this questioning "the problem that has no name." Indeed, my positionality even resembled those of the privileged white women about whom Friedan wrote. Like her subjects, I felt a palpable sense of second-class citizenship and gender inequity, most notably around issues related to the rather restrictive options I saw in pursuing an academic career. I suppose you might say that this was a case of status discontent—my good education and high aspirations coming face-to-face with the realities of the American working world and its lack of flexibility related to issues of gender, especially regarding the choice of motherhood. Up to that time, I had been able to accomplish what I had set out to do. The problem of combining family and work seemed a challenge of a different order.

Dealing with these challenges occurred against the backdrop of a particular historical moment imbricated by the civil rights, antiwar, and women's movements. Together, these provided vocabulary, frame of reference, and a toolkit for addressing what many women experienced as social injustice. Rather than an acute conversion due to a feminist consciousness raising group, abortion, or divorce, it was reflection on my life history that led to a shifting epistemological and normative orientation. The more I studied, lived, and analyzed, the more I became committed to a set of understandings about the gender relations of American society and its inequities, even for relatively privileged women like me, a white, middle-class professor. These epiphanies caused me to question what I knew, how I knew it, and how I wanted to define my academic focus.

As I became more committed to becoming a feminist educator, I used my background in American Studies to embark upon a career teaching, researching, and writing about women's issues from the standpoint of history and the social sciences. My education honed intellectual tools helpful to interrogating the ways in which the social order was arranged, especially regarding what we called in the 1970s and 1980s "sex equity." I became convinced that the ways in which we thought we knew what we knew and judged what we took to be "excellence" were all profoundly shaped by who we were and where we'd been.

When my family moved to Texas in 1978, I found part-time work at a local community college teaching American history. Even more remote from my kinship networks, I wondered again and again why it seemed so hard to combine an academic career with motherhood; why good childcare was so very hard to find and so expensive when it could be found; and why more part-time, flex-time, and/or job-sharing options were not offered by American educational institutions and businesses. There seemed little support in the popular culture, as far as I could tell, for women who wanted to work full-time while raising a family, except of the platitudinous, nonsensical sort epitomized by the media's claim that "women could have it all."

I recognize here that my disgruntlement is that of a woman marked by race and class, realizing full well that "women have always worked" if they are poor or working class, and especially if they are African American. I studied the problem, and found that some European and Scandinavian countries approached women's childbearing and working lives from a more supportive standpoint. Once again, I saw the role that cultural worldview played in shaping structures, institutions, and practices unconducive to women's involvement in this nation's professions.

I began to teach American history in ways that I hadn't been taught myself. I made explicit connections between the situation of women and the civil rights movement, in which my family had participated in a small and localized way during the 1960s. I sought American history textbooks that incorporated more social, cultural, and women's history into the classes I taught. I developed an approach that was very much in sync with the emerging disciplinary framework of the day—history from the bottom up. I had my students conduct oral histories and family histories; I asked questions and "talked back" to the dominant portrayals of progress in the unfolding of our textbook's celebratory nationalistic history.

As a popular contemporary anthology in women's studies puts it today, women's studies resulted in transforming the academy: "People began questioning the nature of knowledge, how knowledge is produced, and the applications and consequences of knowledge in wider society" (Shaw and Lee, 2007). As I've noted here, making this epistemological shift began for me in the 1970s, but it surely gathered steam in the 1980s.

The 1980s

In 1981, upon returning to the northeast pregnant with my third child, I eventually decided that one way to combine motherhood and a teaching career might be to teach at the high school from which I graduated. The institution had a coeducational lower school, with the added bonus of an all-day kindergarten. Moreover, it offered substantial discounts on the cost of tuition to its faculty!

According to traditional valuations of academic status, this new position represented a step down from work in colleges and universities. Feminist thinking had, however, encouraged us to challenge the conventional status hierarchies. At the time, I joked that I was the only working mother who spent more time with her children than mothers who stayed home with their kids. We drove back and forth to school together each day, and, if I needed to, I could see them in the cafeteria at lunch time.

In many ways, the eight years spent teaching at this women's high school proved the fulcrum for a significant shift in my academic career. Moving into high school teaching, especially at a place where I was able to design and teach courses in women's history, not only allowed me to develop my understanding of women's studies more broadly but also provided new opportunities for doing so with a group of like-minded women. My participation in the SEED (Seeking Educational Equity and Diversity) Seminars, led by Peggy McIntosh and Emily Style, was extraordinarily satisfying as well as challenging, in ways that contributed to understanding the limitations of my original understanding of women's studies. In the end, my experience teaching high school made possible a significant redirection of my career in the 1990s.

In response to a year spent discussing on a monthly basis gender and multicultural issues with McIntosh, Style, and my SEED colleagues, I was asked by Oak Knoll's head of school to lead a similar seminar for K-12 faculty interested in considering how to "gender balance the curriculum." At the end of the year's seminar, we held a national conference at which both McIntosh and Style, along with a number of our own faculty and invited guests, addressed this issue. We dealt with issues of gender, race, and sexuality at the conference and published a small monograph of the proceedings called *Listening for All Voices: Gender Balancing the School Curriculum* (Crocco, 1988). With foundation funding, we distributed five thousand copies free of charge to educators across the country.

Style's essay (1988) on curriculum "as window and mirror," first published in *Listening for All Voices*, contributed a rich and expressive metaphor for thinking about the work of change that feminist teachers and scholars were advocating. Later, I would make frequent use of this publication in my work as a teacher educator.

Throughout the year we spent talking about gender, we wrestled with a host of issues—academic, theoretical, and practical. Once we had been shaken from our traditional academic moorings, we struggled to devise new ways of thinking about excellence. In considering the feminist movement and its many public faces, we addressed questions raised decades earlier in the twentieth century by the intellectual adversaries, historian Mary Ritter Beard and Bryn Mawr's first female president, M. Carey Thomas: What does feminism mean? Is it simply access to the male-defined world, as Thomas saw it? Or does feminism mean radically altering notions of what should be valued and how we value it, as Beard saw it? The ways in which feminist scholars split throughout the century over questions related to this divide signaled the emergence of multiple forms of feminism as women's studies grew in the 1980s.

At Oak Knoll, such questions came to a head in our reading of Nel Noddings' (1984) *Caring*. We had a lively discussion about whether her ideas represented a new valuation of women's allegedly traditional orientation to life—caring—or whether adoption of this ideal for teaching simply reinforced stereotypical notions about being a woman. Wasn't the feminist project directed toward uncoupling such stereotypical attributes from our ideas of this gender role? At least for this group of mostly comfortable, mostly white middle-class teachers and mothers, we acknowledged the risks, but found the book's approach congenial. As Peggy McIntosh often said, it had been women's work historically to do the "making and the mending of the social order." Noddings affirmed that this life-sustaining activity had value. The problem lay not in the call to make caring the foundation for teaching, but in a society too blind to recognize the value in such an approach. Since a caring approach was very much in keeping with the progressive educational orientation of a school devoted to the "whole child," most of us felt the book articulated our approach to teaching quite well.

My time at Oak Knoll developed my passion for schools, education, and curriculum development, all of which would be given fuller scope when I moved to Teachers College, Columbia University, in the fall of 1993. Taking up the work of an educational researcher and teacher educator offered fresh opportunities for considering how gender was situated in the field of education. What I discovered about the world of education in general and social studies in particular came as a surprise.

The 1990s

By all accounts, the world of education seems a very female world, yet there is a "missing discourse of gender" found there, according to a report by the American Association of Colleges of Teacher Education (AACTE) (Blackwell, Applegate, Earley, and Tarule, 2000). Much the same could be said of social studies. In the early 1990s, the field's tenured professors, prominent scholars, and executive officers of the

College and University Faculty Assembly of the National Council for the Social Studies were still largely male. Indeed, there had not yet been a single female editor of the field's foremost journal, *Theory and Research in Social Education*. I had surely expected the opposite when I became a social studies professor in 1993.

Since then, I have devoted a good deal of my scholarship and teaching to trying to understand why social studies education paid so little attention to gender, in particular, and diversity, in general. As I explored the history, traditions, and contemporary practice of the field, I found there had been significant feminist activity during the heyday of the women's movement—the 1970s and 1980s. But such efforts had receded, despite the fact that there were more and more women taking their place within the professoriate. In the end, I concluded that multicultural issues had superseded feminist and gender-related considerations in the 1990s, in part, I hypothesized, because of a sense that the inequities women faced had been addressed (Crocco, 2003/2004). I believe this viewpoint reflects the "epistemology of ignorance" (Mills, 1997) about gender (and race) that is all too common in our society, reflecting white, male privilege related to remaining uninformed about matters of consequence to particular "others."

I introduced two new courses to the Program in Social Studies at Teachers College: "Diversity and the Social Studies Curriculum" and "Women of the World: Issues in Teaching." In both courses, students consider issues related to the transformation of knowledge (Minnich, 1993) and curriculum (Banks, 1993; McIntosh, 1983, 1990). Although such courses have become commonplace in schools of education during the last 15 years, addressing gender and sexuality in teacher education is relatively rare even today in many schools of education.

When a letter of invitation for a program on gender and education arrived from Ellen Silber of Marymount College in Tarrytown, NY, a few years later, I applied immediately. This Ford Foundation-sponsored project began from the premise that there was still much work to be done on gender in this female-dominated field; indeed research conducted under the auspices of the project underscored the findings of the AACTE study concerning the "missing discourse" of gender in teacher education. Subsequent research by Karen Zittleman and David Sadker (2002) on teacher education texts confirmed this deplorable state of affairs.

In social studies, the issues are related, although distinct. For example, it seems that the kinds of courses often taken by those preparing to be history and social sciences teachers are very traditional ones. I have met certified teachers who have taken introductory-level survey courses in history as part of their undergraduate teacher preparation work that paid little attention to women's history. I have found that my own students, who typically graduated from college with undergraduate history majors, fall into two camps. For some, notions of knowledge transformation are old hat; for others, these ideas are entirely new. Indeed, I've been chagrined to find so many practicing teachers who, even today, approach curriculum, curriculum standards, and high-stakes testing as valid representations of the knowledge with greatest worth in our society. They may chafe under the pressures of test prep routines, to be sure, but few question the fact that the New York State Regents exams in U.S. History include virtually no questions of any substance about women. Most assuredly, the standards and accountability movement in education has shrunk the available space for dealing with women as forms of "traditional" curriculum in history get recirculated. By contrast, Advanced Placement exams given in high schools for college credit today include more and more social history, including women's history. In fact, in many ways, the school-based standards and accountability movement can be read as a backlash against newer forms of knowledge—of both a gendered and multicultural type.

Interestingly, feminist scholarship of the last 20 years has questioned the utility of the concepts of "woman" and "gender" (e.g., Butler, 1990; DeLauretis, 1984; Scott, 1999), reflecting the tension between unity and difference in doing women's studies. In my classes, we spend time addressing a few of these theoretical issues, especially essentialism, patriarchy, and gender socialization. We examine the place of women in American and world history and explore contemporary issues related to women's status worldwide. As global feminisms have arisen, we talk about the ways in which the priorities and politics of Western women intersect and divide from those of women around the world. Increasingly we discuss issues of masculinity and sexuality as part of these conversations. These theoretical considerations continue to stretch my epistemological orientation in ways that I could not have predicted 20 years ago. Although I find them fascinating, the work of teacher education tilts in an applied direction. My teaching is very much focused on the work of gender balancing teacher education curriculum and understanding the ways in which gender continues to shape the "commonplaces" of schooling (Crocco, 2006) and social studies research engages (or not) with gender (Hahn, Bernard-Powers, Crocco, and Woyshner, 2007).

In the summer of 2007, I introduced an online version of "Women of the World." One of the assignments required students to look at the Web site devoted to Judy Chicago's "The Dinner Party," which became the centerpiece of the Brooklyn Museum's new Elizabeth A. Sackler Gallery for Feminist Art when it opened in March 2007. Students were to take a close look at the representations of the 39 women with a place at the table and the 999 women represented on the floor tiles of the exhibit. Their task was to consider how they might use this exhibit in their teaching. Many students had never heard of this iconic work, but were intrigued and, in some cases, astounded by it. The place of honor Chicago's work had achieved at this prestigious museum seemed a positive sign of where women's studies stood in the early twenty-first century. Likewise, the opportunity to use the Internet as a means of exploring women's history collectively, with an online class comprised of students from New York, San Francisco, Houston, and Singapore, suggests the extraordinary possibilities for "social networking" this medium provides (for an extended discussion of this topic, see Crocco, Cramer and Meier, 2008)

Conclusion

Bringing a women's studies perspective into social studies education has meant concentrating my work as teacher and researcher on the longstanding tradition of using women's studies as an agent of change. Many scholars have written about the political project of feminism—to change the academy and change the world. As the number of women in social studies continues to climb, they have achieved some notable firsts, including first female editor of *Theory and Research in Social Education* in 2001. Likewise, many leading researchers today are women. Even the recent review of research on gender and social studies education (Hahn et al., 2007) demonstrates a fair amount of activity despite the bleaker picture found in social studies textbooks, according to recent publications (Clark, Allard, and Mahoney, 2004; Clark et al., 2005). Such research affirms how little progress has been made, at least as measured by "official" versions of history found in textbooks. Perhaps this should come as no surprise, given the conservative climate of schooling and the nation over the last 20 years. Looking forward, social studies educators at all levels will need to be mindful of their own epistemological and personal journeys in order to commit themselves to improving the visibility of women and gender in our field.

References

Banks, J. (1993). *An introduction to multicultural education.* Boston: Allyn and Bacon.

Blackwell, P., J. Applegate, P. Earley, J.M. Tarule. (2000). *Education reform and teacher education: The missing discourse of gender.* Washington, DC: American Association of Colleges of Teacher Education.

Butler, J. (1990). *Gender trouble: Feminism and the subversion of identity.* New York: Routledge.

Clark, R., J. Allard, and T. Mahoney. (2004). "How much of the sky? Women in American high school history textbooks from the 1960s, 1980s, and 1990s." *Social Education*, 68(1): 57–62.

Clark, R., K. Ayton, N. Frechette, and P. J. Keller. (2005). "Women of the world, re-write! Women in American world history high school textbooks from the 1960s, 1980s, and 1990s." *Social Education*, 69(1): 41–44.

Crocco, M.S. (2006). "Gender and social education: What's the problem?" In E. Wayne Ross (Ed.), *The social studies curriculum* (3rd ed.). Albany, NY: State University of New York Press, 171–193.

———. (2003/2004). "Dealing with difference in the social studies: A historical perspective." *International Journal of Social Education*, 18(2): 106–126.

———. (1988) *Listening for all voices: Gender balancing the school curriculum.* Oak Knoll School: Primerica Foundation.

Crocco, M.S., J. Cramer, and E.B. Meier. (2008). "(Never) mind the gap! Gender equity in social studies research on technology in the 21st century." *Multicultural Education & Technology Journal*, 2(1): 19–37.

DeLauretis, T. (1984). *Alice doesn't: Feminism, semiotics, and cinema.* Bloomington, Indiana: Indiana University Press.

Friedan, B. (1963). *The feminine mystique.* New York: Norton.

Hahn, C., J. Bernard-Powers, M.S. Crocco, and C. Woyshner. (2007). "Gender equity and social studies." In S. Klein (Ed.), *Handbook of research on gender and education.* Mahwah, NJ: Lawrence Erlbaum, 335–359.

Kuhn, T. (1962). *Structure of scientific revolutions.* Chicago: University of Chicago Press.

McIntosh, P. (1983). "Interactive phases of curricular re-vision: A feminist perspective." Working Paper No. 124. Wellesley, MA: Wellesley College.

———. (1990). "Interactive phases of curricular and personal re-vision with regards to race." Working Paper No. 219. Wellesley, MA: Wellesley College.

Mills, C. (1997). *The racial contract.* Ithaca, NY: Cornell University Press.

Minnich, E. (1993). *Transforming knowledge.* Philadelphia, PA: Temple University Press.

Noddings, N. (1984). *Caring.* Berkeley, CA: University of California Press.

Sadker, D. and E.S. Silber. (2007). *Gender in the classroom: Foundations, skills, methods and strategies across the curriculum.* Mahwah, NJ: Lawrence Erlbaum.

Scott, J.W. (1999). *Gender and the politics of history* (rev. ed.). New York: Columbia University Press.

Shaw, S.M. and J. Lee, J. (2007). *Women's voices, feminist visions: Classic and contemporary readings* (3rd ed.). New York: McGraw Hill.

Style, E. (1988). "Curriculum as window and mirror." In M.S. Crocco (Ed.), *Listening for all voices: Gender balancing the school curriculum.* Oak Knoll School: Primerica Foundation, 6–12.

Women's Studies Department, University of Maryland, www.womenstudies.umd.edu/about/missionstatement.shtml accessed on October 8, 2007.

Zittleman, K. and D. Sadker. (2002). "Gender bias in teacher education texts: New (and old) lessons." *Journal of Teacher Education*, 53(2): 169–180.

2

Gender at the Center:
The Making of an Educator

Joan Poliner Shapiro

Introduction: Two Critical Incidences

In this chapter, I will carry out a considerable amount of introspection. This reflective thinking has come after almost 40 years in the Academy. My process will focus on issues related to Women's Studies that include: feminism; politics in the field; interdisciplinary studies; students, content, and pedagogy in this area; special evaluation and assessment; networking; and Gender Studies. However, before I turn to these topics, I want to describe two critical incidences that profoundly affected me and helped me focus on gender as my most important category of difference.

My awakening to the issue of gender did not come to me at an early age. In fact, it took a couple of critical incidents to bring it to the forefront. The first incident occurred at Uncle Max's funeral while I was in my mid-30s. This event took place in the northern part of England, in which a very fundamentalist sect of Jews lived. When my husband and I arrived at Uncle Max's home, the women were moaning and wailing around a hearse that waited outside the door. This seemed strange to me as Uncle Max was well into his eighties and had not suffered unduly before his death. I went to the burial ground for the ceremony with my husband. At the grounds, much to me surprise, I turned out to be one of two women present and we were both told not to leave our cars. Apparently, women were not allowed on the burial group lest they "sully the soil." This was a painful experience for me as I had only recently buried my father in the conservative Jewish tradition, and my mother, sister, daughter, and I had been free to mourn in public and on the cemetery grounds. It seemed to me that the humiliation for women continued that day when the Rabbi told Auntie Minnie, Uncle Max's wife of 45 years, that she missed an excellent speech he had given on behalf of her husband on the burial grounds. All the women seemed to accept what I perceived to be an insult without comment, but I was never the same. The category of gender emerged as the most salient aspect of diversity in my life because of this critical incident (Shapiro, Sewell, and DuCette, 1995, xvi–xvii).

Another incident, making certain that I would focus on gender, occurred when I had just finished my doctoral work and was seeking employment as an assistant principal in suburban districts. I had received a number of negative replies to my applications and wondered why this was so. After all, I had received my doctorate from the University of Pennsylvania and had even finished a postdoctoral experience at the University of London. "Why was I unemployable?" I asked some friends, while I was visiting my alma mater. A female professor, the only one in the graduate school, overheard my lament. Although I had never had her as an instructor, she asked me to stop by her office. I did so, and she asked the following question: "Are you a teacher in the School District of Philadelphia?" I replied, "No." She informed me that my chances to become a principal in the area, then, were negligible. The professor indicated that female Philadelphia employees stood a slim chance to become principals in the district, but those outside, in the suburbs, stood no chance. All of this was extremely hard to hear. But thanks to this professor, who turned out to be an unexpected mentor, I decided to make a career change and began looking for employment in higher education. This decision turned out to be a wonderful one for my future and I never looked back once employed in the Academy (Shapiro et al., 1997, 173–174).

Evolution of a Field and a Feminist

Over the years, I have had diverse experiences in higher education that have changed my perspective on feminism. For example, moving from the codirecting of a women's studies program in the Graduate School of Arts and Sciences to a College of Education where I teach a gender course has changed my definition of Women's Studies' goals and purposes. While helping to lead a program, Women's Studies was at the very center of my professional life. I placed women's and girls' needs as a top priority and thought about how to provide primarily my female students with the kind of knowledge base they needed to succeed in their professional and even personal lives. Arriving at Temple University's College of Education, I have broadened my focus to include all students, both women and men. It does not mean that I have turned my back on Women's Studies. In fact, for many years I was on the steering committee of the program at Temple. However, while thinking about the university and the needs of women, at the College of Education, I have tried to add to the knowledge base of all my students by designing and teaching Gender Issues in Education. Interestingly enough, I had trouble getting approval for this course. Some of my male colleagues thought there would not be enough students to take it and that there was no need for it. But their beliefs proved to be wrong as the course is always well enrolled with usually 25 percent men and the rest women. In fact, it generally is a course with a diverse composition attracting students throughout the College. A year after it was passed, a new African American colleague, James Earl Davis, arrived at Temple who genuinely wanted to teach the course as his scholarship centered on African American boys. Frequently, we team-teach or take turns instructing it. So, my senior colleagues had been proved very wrong indeed.

But there is something about the positionality in which one exists. Today, for instance, I am in a College of Education and tend to think of gender as an important area for all my students. However, recently I have begun to feel ambivalent because I am serving on the Provost's Academic Planning Steering Committee for the whole university. In this role, I find myself thinking in an interdisciplinary fashion once again and am trying to take into account ways in which more students throughout the university can take Women's Studies and even gender studies courses.

The Politics of Women's Studies

As politics changes, so does Women's Studies. What I mean by this is that in certain eras I have tended to see my role as radical and even subversive, but in other time periods less so. For example, when I was team-teaching with Carroll Smith-Rosenberg, a course called Women and Men and the Ethical Crisis of the 1980s to Penn students, we were living within a Reagan administration. This era made it essential to speak out, yet always be mindful that in Washington, DC, a very conservative group of individuals held sway. When Carroll and I spoke of welfare and medical health insurance for all Americans, one student asked us to be patient with her when dealing with these kinds of concepts. After all, she said, as far as she could remember, Ronald Reagan had always been president of the United States. It is clear when we taught that our message might have sounded very radical to some of our students, but it was not really that outrageous. In an era of Jimmy Carter or Bill Clinton, what we have had to say in Women's Studies does not seem to be very radical. With George W. Bush in the White House, once again, we sound subversive. So each administration makes us appear radical or subversive or not.

I have always thought of Women's Studies as the academic wing of the women's liberation movement. To me, academics and activism are entwined but not identical. On the Penn campus, we had a Women's Center that in many ways focused on aspects of women's liberation. They organized the marches, sit-ins, and other operations. Nevertheless, in Women's Studies, we took part in these actions, sometimes in the vanguard and sometimes joining the ranks.

However, an activity that we did try to lead, when I worked in the area of Women's Studies, was to add one diversity course to the undergraduate curriculum. That could include courses in African American Studies, Chicano Studies, and other diverse programs and departments. However, the disciplines fought against us. They did not want to give up any part of the curriculum to an interdisciplinary, radical program. They battled hard and won.

Going beyond the Academy, in the area of politics, there are some issues that never seem to fade away. They require determination and dedication lest we slip back into an abyss. Each new generation needs a good dose of feminist history as well as frequent reminders about the need for the following: pro-choice, gender equity, a strong Title IX, more women in science and math, and respect for diversity.

Women's Studies: Interdisciplinary/Disciplinary

There is a paradox concerning whether Women's Studies should be on the one hand, a separate discipline while on the other, a program or department retaining its inherently interdisciplinary nature. This paradox has existed from the beginning, when Women's Studies Programs first entered the Academy. For example, in the 1980s, when I was codirector of Women's Studies, I thought it should be a separate discipline and that we should have some standing faculty to represent us within the institution. In my current role, as a professor in a College of Education, I am pleased that it is interdisciplinary in nature and that it continues on as a program in my university, although it now has some core faculty members. However, because of its interdisciplinary nature, I have to admit that I am troubled that neither my colleagues nor I in education can direct the program as we are in the wrong school. The College of Liberal Arts is the one in which Women's Studies is housed and in which directors are selected. If Women's Studies is a truly interdisciplinary program, then this should not occur. So, for me, it is positionality or our standpoint as to what we think it should be. What I mean by this is that it may depend on your position in the institution to determine what kind of program or department Women's Studies should become.

Students, Content, and Pedagogy in Women's Studies and Gender Studies

Overtime, there has been an ebb and flow of conservatism in Women's Studies and gender studies programs. Of late, thanks to David Horowitz (2006) and other critics, I am more careful about what I say in class. This is sad, as all too often free speech, which is the hallmark of the Academy, is not operating as effectively as it should. I know that I have changed in the way that I present myself because when I worked some years ago with my colleague, Jackie Stefkovich, on ethics, I wrote that I presented where I stood on issues to my class. However, I made it clear to my students that I would value their views and perspectives. Jackie, on the other hand, who had been trained as an attorney, tended to behave in a more neutral, unbiased way (Shapiro and Stefkovich, 2001, 2005). Now, I hold back a bit more until I take the temperature of the class and determine if it is safe to present to them who I am and where I stand on issues.

Not only have I changed in my forthrightness with my classes, but over the years, I have modified to a certain degree about what I ask students to read. However, there are certain books and papers that I constantly turn to. For example, I have them read works by Charlotte Perkins Gilman, Carroll Smith-Rosenberg, Carol Gilligan, Belenky et al., Nel Noddings, and other classical selections. But I also add the men's and boy's literature and writings in leadership as well as cross-cutting works in race, social class, and gender. The field is so large now in Gender Studies, that although I assign a number of readings including my own co-authored text (Ginsberg, Shapiro, and Brown, 2004), in the second half of the class, students introduce us to current readings. They do this through Journal Club where they share journal articles or book chapters that follow their interests in the area. Students are asked to discuss the background of the authors and lead a discussion. I believe strongly in constructivist education where students help to develop the class work. Journal Club provides an outlet for them in designing the course. Thus, each time I teach Gender Issues in Education, it is fresh and stimulating.

Along with a constructivist approach, pedagogically I utilize connected knowing (Goldberger et al., 1996). To develop this theory within the classroom, I try to turn first to autobiography because the students need to know themselves and reflect upon what they value before understanding and connecting to others. We then move toward dealing with difference. The differences encompass not just gender, race, and social class, but also deal with disabilities, age, transgender, and diverse learning styles. Because of defining diversity widely, the spectrum has broadened. Students are not afraid to talk and write about their thoughts and issues. For example, most recently, after taking part in an activity involving silence and voice, here is what an African American female student wrote:

> I absolutely love your class, it reminds me of my Women Studies classes (the good parts of course), and it has surprised me that I am even excited about the different perspectives in class. I am used to sitting in classes with progressive (or non progressive) Feminist, Womanist, Black Nationalist, Afrocentrist, etc. However, this class is full of professionals—teachers and principals. It's quite interesting to say the least. At first I felt indifferent to speak so much about gender or race or gender AND race, yet I feel as if a space has opened up to do so.

Journal writing has been an invaluable assignment for a class that focuses on gender and connected knowing. Carroll Smith-Rosenberg and I used journals as a way to make certain that students not only read the assigned readings but also developed a critical stance about what they read. Through a study of their journal entries, we were able to determine the writings that proved to be meaningful to them. Additionally, the journal reflected a student's growth or the lack of it overtime. For example, after reading *The*

Yellow Wallpaper, by Charlotte Perkins Gilman (1973), a first year undergraduate at Penn wrote:

> O.k. I just finished *The Yellow Wallpaper*. We discussed it in class. I'm not sure if I like the book or the class. I want to be honest and I don't want a bad grade. Well, I don't have to turn this in. I was excited to take a Women's Studies class because I want to discuss and learn things about culture—my culture, and about me. I also want to see the different perspectives between east Philadelphia and West L.A. *The Yellow Wallpaper* first definitely was one of the weirdest books I've read. I would have it would remind me of I Never Promised You a Rose Garden or maybe Sybil, but it didn't really. First I was struck by the lack of assertiveness she had. How could he ALWAYS be right? How frustrating! And the Dr. didn't seem to help—obviously since she went a little bonkers. Also, the description was intense! I could see her seeing this lady, and I believe it was there (Shapiro and Smith-Rosenberg, 1989, 205).

As the course continued, this same young woman, who previously had no sense of the history of women, began to change her perspective. In another entry, in which *The Yellow Wallpaper* was not part of the readings, the student stated:

> In class, we talked about a basis for our society and things going on now. The deal with the moral majority wanting complete control over their children, a new Conservative view and how this is or can be stemming from the Greek culture where the father actually owned his children—*The Yellow Wallpaper* fits in, I gather, because of what a the woman kept saying about her being inferior. Her husband has control, respect, and success. He treats her like a child. As you watch her go mad your emotions get involved. It's horrifying to read about her perceptions changing and her pain. Her confusion and oppression scare me. Is our society so different? Are our society's expectations of gender roles and power relations so different? Maybe I didn't like the class because suddenly I'm looking at things I don't want to see. I see injustice, pain, etc. in *The Yellow Wallpaper*, but that is different from my life. I do want to see how this course affects my life and if it does. This woman was a good education (Shapiro and Smith-Rosenberg, 1989, 205–206).

These two passages demonstrated that the student's view changed through a better understanding of white middle- and upper-class women's lives in the 1890s in the United States compared to her own contemporary life. The historical perspective helped her to grow and reach more of a connection with her own life as well as that of the past.

Evaluation and Assessment in Women's Studies

Overtime, I began to realize that Women's Studies Programs deserved their own special kind of evaluation and assessment in line with their underlying philosophy. By this I mean that this area does not pretend to be neutral, unbiased, and remote. Instead, there are some deeply held perspectives and beliefs that are part of these programs and deserve to be recognized. For example, cooperation over competition is valued in this field and engagement of all participants is expected. Thus, I began to develop a new approach that I called participatory evaluation (Shapiro, 1988). The expectation for this kind of review is that the evaluator is part of the process from the beginning of its development until the end of the funded project and sometimes even beyond this point. The evaluator is not silent in the process and also makes certain that all those who wish to speak have a voice as well. The focus of these evaluations tends to be on formative rather than summative assessment. In this way, the project moves along from site to site or from its inception to its end with constant feedback from the evaluator. Consistently, because of this participatory approach, there is an openness to collaborations, allowing me to write articles and even books with program directors, facilitators, and others who designed and implemented the project.

The Next Move serves as a good example of how this kind of participatory evaluation worked. This project was designed by Higher Education Resources (HERS)/Mid-Atlantic, and I served as the evaluator, helping to refine the project as it moved along from site to site. The concept behind the project was that women in higher education needed some staff development to help them move up the postsecondary hierarchy. Observing a group of secretaries, led by a vice provost, I noticed that a few of the participants were silent. At the end of the session, I spoke with the vice provost and asked her why she was chosen to lead the group. She said that she had an eye on her secretaries and helped them surmount barriers to move up. I then interviewed a silent secretary in the group, who turned out to be the vice provost's assistant. She said that she did not know why her superior was chosen to head the group. She said that the vice provost never said "hello" to her in the morning. In fact, she seemed not to know that she existed.

After reporting on these interviews to the sponsors, they decided to choose administrators who had work experience as secretaries at the next site (Shapiro, Secor, and Butchart, 1983). Hence, formative evaluation had been helpful in refining the original project.

In the Gender Awareness through Education (GATE) project, I assisted the project director and facilitators to help improve the interventions of the various groups that met throughout the three-year period of the grant. This role enabled me to become in many ways part of the project. Eventually, it led to a book being written by the project director, a facilitator, and me (Ginsberg, Shapiro, and Brown, 2004). Because of our different roles, we were able to provide insights into this project from diverse perspectives.

So, to summarize, to me evaluation or assessment is very much part of the review process and far from remote and neutral. This kind of participatory approach seems to dovetail much more with the philosophy of Women's Studies or Gender Studies programs, enabling changes to be made in the project and allowing for all kinds of collaborations between those being evaluated and the evaluator.

Moving Across the University and Networking

At the University of Pennsylvania, I did work with African American Studies, and various traditional departments and colleges, including sociology, history, and the Graduate School of Education, although it was never easy to accomplish. There has always been a budgeting problem that tended to put up barriers between programs and department and schools. Now, in the College of Education, while I do work with others who are dealing with differences, I have not worked across schools or colleges. However, I do team-teach quite often with an African American scholar, James Earl Davis, who has studied in both African American Studies and Women's Studies programs. Unfortunately, the budget tends to be very much of an obstacle. Whether enrollment based budgeting or responsibility based budgeting, barriers are put up that make it challenging to work with others in the university.

Networking has been invaluable to me for my own growth as a scholar. At one stage in my life, I was part of a feminist reading group that included faculty from universities and colleges all over Greater Philadelphia. I have also been a part of numerous grants, frequently serving as an evaluator, and through these projects, I have met wonderful colleagues. Most recently, through an educational movement that I spearheaded with my colleague, Steve Gross, called The New DEEL (Democratic Ethical Educational Leadership), my network has truly expanded. I am working with colleagues in more than 20 universities and with practitioners in Canada, the U.K., Australia, and Taiwan. I've never attempted to start a movement of my own, but it has proven to be extremely empowering. Hopefully, the New DEEL will help educators emphasize the development of wise, intelligent, and ethical future citizens instead of focusing on students' test results and accountability issues.

Conclusion: Women's Studies or Gender Studies?

By now you no doubt have surmised that I would much prefer the term Gender Studies over Women's Studies. More and more I feel that boys and men as well as girls and women need a feminist education. If one-half of the population is left out of learning about the other, it cannot be a benefit to society. There is so much that we have learned about sex and gender that needs to be shared. The only way to do this is by reaching out to both groups.

I saw a fine example of reaching out to both males and females some years ago while I served as an evaluator on a project called *The Courage to Question: Women's Studies and Student Learning* (Musil, 1992). One of my tasks was to review the a program at Lewis and Clark College. I arrived at just the right time. The Gender Studies Program was sponsoring a conference in which undergraduates presented their work on gender. It seemed to me as if the whole student population was involved in this experience. Michael Kimmel was the keynote speaker, and males and females had a voice during this event. That exciting conference left an indelible impression on me. I could see how inclusive a Gender Studies Program could be.

When I first entered this field, I thought I was a left-wing humanist. Then, as I worked in Women's Studies, I became a liberal democratic feminist. Now, in a College of Education, I believe that I am a humanist again. I feel that all voices should be heard. Boys and men deserve attention as well as girls and women. So, it seems to me, after almost 40 years in the Academy, that I have come full circle. That does not mean I will not continue asking the question: What about the girls? But I don't think I will forget: What about the boys? As a granny of a boy and a girl, I feel strongly that both deserve the best possible education. Without a strong and sustained knowledge and understanding of gender, they will not receive it.

References

Belenky, J., B. Clinchy, N. Goldberger, and J. Tarule. (1986). *Women's ways of knowing.* New York: Basic Books.

Gilman, C.P. (1973). *The yellow wallpaper, 1892.* New York: Feminist Press.

Ginsberg, A., J.P. Shapiro, and S.P. Brown. (2004). *Gender in urban education: Strategies for student achievement.* Portsmouth, NH: Heinemann.

Goldberger, N., J. Tarule, B. Clinchy, and M. Belenky. (Eds.) (1996). *Knowledge, difference and power.* NY: Basic Books.

Horowitz, D. (2006). *The professors: The 101 most dangerous academics in America.* Washington, DC: Regnery Publishing.

Musil, C.M. (Ed.) (1992). *The courage to question: Women's studies and student learning.* Washington, DC: Association of American Colleges.

Shapiro, J.P. (1988). "Participatory evaluation: Towards a transformation of assessment for women's studies programs and projects." *Educational Evaluation and Policy Analysis,* 10(3): 191–199.

Shapiro, J.P., C. Secor, and A. Butchart. (1983). "Illuminative evaluation: assessment of the transportability of a management training program for women in higher education." *Educational Evaluation and Policy Analysis,* 5(4): 465–471.

Shapiro, J.P. and C. Smith-Rosenberg. (1989). "The 'other voices' in contemporary ethical dilemmas: The value of the new scholarship on women in the teaching of ethics." *Women's Studies International Forum,* 12(2): 199–211.

Shapiro, J.P. and J.A. Stefkovich. (2001). *Ethical leadership and decision making in education: Applying theoretical perspectives to complex dilemmas.* Mahwah, NJ: Lawrence Erlbaum Associates.

———. (2005). *Ethical leadership and decision making in education: Applying theoretical perspectives to complex dilemmas* (2nd ed.). Mahwah, NJ: Lawrence Erlbaum Associates.

Shapiro, J.P., M.B. Kenney, R. Robinson, and P. DeJarnette. (1997). "The 'rite of passage' of non-traditional doctoral students in educational administration: Crossroads in gender, race and social class." *Journal of School Leadership,* 7: 165–193.

Shapiro, J.P., T.E. Sewell, and J.P. DuCette. (1995). *Reframing diversity in education.* Lancaster, PA: Technomic Publishing Company, 1995. Book reprinted in paperback- Lanham, MD: Rowman & Littlefield, 2002.

3

Women Studies—The Early Years: When Sisterhood Was Powerful

Paula Rothenberg

Women's Studies and the Women's Liberation Movement, which for me are almost impossible to separate, have been among the defining elements of my life. They have defined who I am and shaped the choices I have made at almost every step of the way since I first became aware of women's issues. I grew up in a world where all women were second-class citizens, where people of color were denied basic human rights, and where economic issues defined peoples' life chances. As if this wasn't enough, many of us were prevented from recognizing the gravity of our condition by a set of beliefs, stereotypes, and falsehoods that we had learned to internalize from an early age. And even when we did understand something of our subordinate condition, many of us simply blamed ourselves.

Regrettably, this is still true today although the particular ways in which we have come to believe this has changed and continues to change as the force of a still powerful unconscious ideology continues to mislead us about the situation of women in the world. On the one hand, the gains that some of us have made are undeniable and on the other, capitalist patriarchy continues to have an uncanny ability to snatch our defeats from the jaws of our victories. This, of course, is not surprising. Early Women's Studies was rife with possibilities that seemed to hold out the opportunity for real and lasting change. But as those of us who were part of Women's Studies even before it was a discipline came to realize, the discipline was transformed both by the academy as it was institutionalized and professionalized and by corporate culture as it was trivialized. One step backward, two steps forward, and then we go back again. Too many well-intentioned women have been duped by a version of feminism that encourages them to define their interests and adopt a definition of liberation that leaves them less rather than more empowered.

But I am getting ahead of myself, and this is not where I wanted to begin at all. I want to begin, not by talking about the disappointments that have been an inevitable and not unexpected result of creating an entirely new field of study, but about the brief and exhilarating time when everything seemed possible. I want to remember the way it felt to be part of a movement that was bigger than we could have dreamed of. I want to remember the early years that brought with them unimaginable opportunities and excitement, the promise that nothing would ever be the same, the sense that a real revolution was occurring and that many of us were at the heart of it. I want to capture the sense of what it was like when I began teaching Women Studies in the early 1970s, those years when we were drunk with excitement and possibilities and when for a brief time, the world seemed to be ours. I want to think about what Women's Studies meant to so many of us during those early years. Those were heady times and it is important not to forget them.

The Personal Is Political

It is almost impossible to convey what it meant to us to teach about women's lives and women's history for the first time. For many of us, Simone De Beauvoir's *The Second Sex* introduced us to an entirely new and shocking picture of the world and our place in it. Within a decade or two, this account of women's place in society was reinforced and translated into the American idiom by Betty Friedan's *The Feminine Mystic* and Kate Millet's *Sexual Politics*. Thinking back, I wonder how it was possible for us to have failed to see ourselves and our lives so completely for so long.

The overriding question that so many asked was, "Where are the women?" We discovered ourselves as we discovered a past that had been hidden from so many, not just women—but the lives of most men as well. At first, we searched for our heroines, the women who we knew must be there but who had been denied their voice, their achievements, their presence. Lost artists, and scientists, and writers and historians and poets and warriors. But soon this compensatory history, important and empowering as it was, was not enough and we realized that when asked, "Where are the women?" we needed an answer that dug deeper into the past and that provided us with a perspective that allowed us to understand how profound that question was and how broadly it needed to be interpreted. That simple but earth-shattering perspective was the recognition that the personal is political. That insight changed everything. What previously had been dismissed as *merely* personal or trivial, and therefore unimportant, was suddenly seen in a new light.

The phrase "the personal is political," first appeared in a piece by Carol Hanish that was edited by Shulamith Firestone and Anne Koedt and that appeared in *Notes for the Second Year: Women's Liberation*. Over the years, that phrase has meant many different things, but at least at the start, it offered a new and radically different way of thinking about reality. Many of us had been taught that genuine scholarship was objective and neutral and that the mark of a true scholar could be summed up by the compliment several of my philosophy students gave me one day when they assured me, not without admiration, that "we don't know what you think about anything!" Fortunately, this ignorance was short lived and we rapidly discovered that it's neutrality and the perspective from which it proceeded wasn't "neutral" at all. Masquerading as objective, it became clear that even so-called objective science was not value free. At the heart of Women's Studies and framing the perspective from which it proceeds was the critical insight that "the personal is political."

Kate Millet's catalytic book *Sexual Politics* armed us with a new and liberating perspective. Along with de Beauvoir, she urged us to recognize the centrality of the concept of patriarchy to understanding women's place in society. Other feminists and left social critics built on this insight when they wrote a variety of works with titles ranging from "The Function of the Orgasm" to "The Politics of Housework." These two women's movement classics in particular typify the kind of analysis generated by the view that so-called trivial aspects of personal life reflect the fundamental relations of power that extend throughout our lives. Even more to the point, they challenged the whole attempt to portray areas of greatest concern to women as unimportant by pointing out that this view was itself a reflection of the politics of our society.

For many of us who rejected the idea that knowledge could be value free, an accompanying realization was that the alleged split between the personal and the political was neither true nor helpful. Aristotle had pointed out more than 2000 years earlier that issues of power between the sexes are political because in such relationships one rules and the other is ruled. He went on to explain that such relationships are predicated on inequalities of power and that dominance and submission are built into them. To put it another way, such relations involve a struggle for control between individuals of unequal power and status who confront each other with essentially opposed interests. It took Women's Studies and the Marxist-Feminism of the 1960s and 1970s to remind us once again that the split between the personal and the political, between private and public life, between home and work, is itself a reflection of the political needs of capitalism and patriarchy, not some eternal, historical truth of human relations and social organization.

The pervasive nature of such struggles was recognized by the New Left and the Women's Movement of that period insofar as they emphasized the need to minimize the inequalities of power that politicize relationships. Teachers tried hard to equalize relations with students by giving up or sharing the power to grade and determine curriculum. Some faculty and students rejected styles of dress that incorporated the expression/assertion of different social status and power. Some deliberately took the title "Coordinator" rather than Director or Chair (of Women's Studies) so as not to buy into language that deliberately or inadvertently perpetuated hierarchy.

In some areas, these efforts were met with moderate success. But substantial long-term success was rare. This was because personal struggle carried out on an individual basis can only have limited success. These relationships, no matter how seemingly "personal" they are, are part of the broader fabric of society and create and maintain inequalities of power and status. This new beginning, the awareness that the personal was political, framed the way many feminist scholars chose to approach their scholarship and shaped the way transforming the curriculum was approached by many of us in the academy. Later it was itself transformed by multiculturalism and was incorporated into both disciplines as the recognition that knowledge in positional. In other words, that one's position or place shapes the way we see the world.

Transforming Knowledge

The first Women's Studies Department in the nation was founded at San Diego State University in the early 1970s. It grew out of the extraordinary activism and energy of the Women's Liberation Movement of the 1960s and was inseparable, at least at the beginning, from the intellectual excitement and political commitment of that time. During the late 1960s many of us became acutely aware of the absence of women and women's perspectives from the intellectual past and present and in seeking to redress this absence, we brought not only an intellectual commitment to remedying it but a passion to the commitment that in my estimation has never been matched since. This was not just about adding women to the curriculum; it was about radically transforming the way knowledge had been constructed. It was about changing the way we thought about ourselves and the world. It was about rethinking everything! As Catherine Stimpson has written in her *Women's Studies in the United States*, A Report to the Ford Foundation published in 1986, "Political activism on campus and curriculum reform were only the most obvious manifestations of the revolution in American high education in the 1960s...By the 1960 the university had also become an instrument for social change dedicated to providing access and social justice...Feminists offered scholars an agenda for research, while scholars provided activists a theoretical framework and data to form the basis for social policy and progress" (Stimpson, 13). She went on to explain: "a feminist perspective informed the work of Women's Studies practitioners, asking questions that others have ignored. The simplest of these was the question 'Where are the women?' " (Stimpson, 14).

The first Women's Studies Course at my school, Paterson State College in Northern NJ, 40 minutes from mid-town Manhattan and a world away, was offered in 1971 or 1972. Typical of many early Women's Studies courses, it was called "Women in Literature" and was offered by the English department. The course reading list included Kate Chopin's *The Awakening*, Ivy Compton Burnett's *Man and Wives*, Germaine Greer's *The Female Eunich*, Lorraine Hansberry's *A Raisin in the Sun*, Sue Hoffman's *Diary of a Mad Housewife*, Paule Marshalls' *Brown Girl, Brown Stones*, Catherine Bird's *Born Female*, as well as works by May Sarton, Doris Lessing, Mary McCarthy, Edith Wharton, and Virginia Wolf.

These courses gave women an opportunity to look at the lives of women from different backgrounds, races, and classes. According to one syllabus of that time, and fairly typical of its approach, the course was designed to provide a study of women as they saw themselves depicted by men, with an emphasis on the uniqueness of the feminine experience as realized in literature. But instead of stopping there, and this is what differentiated Women's Studies from other fields, the course announced that it would begin with a particular perspective, the assumption that women throughout history had been treated as an oppressed class. The course would explore the roots of their oppression and discuss reactions to it by women writers.

A syllabus of a typical Women in Literature course might have been divided into three parts, the first section typically would have focused on "The Traditional Role of Women: Love, Marriage and Motherhood." Part Two might have looked at "Unhappy Adjustments to Traditional Roles: Marriage and Motherhood, Professions and Creative Expression," while the final section might have imagined what women's writers lives would be like under the heading "Free Women: Evolving Life Styles, Psychic-Economic Independence, Sexual Freedom, Political Radicalism."

I started teaching my first Women's Studies course in the early 1970s. It was called Philosophy of Sexual Politics and it was offered in the Philosophy Department at Paterson State under a special course number designed for courses that would only be offered once or twice on an experimental basis. Sexual Politics? Why would anyone think that the topic had enduring value?

In 1975, Paterson State managed to get a concentration in Women's Studies passed. It consisted of courses in Women's Changing Roles, Women in Literature, Women's Health, and so on, and then in 1977, rather miraculously I think, we were able to pass a minor. I think that was because we were developing Women's Studies at a state college in New Jersey rather than a more prestigious institution, and, to be honest, no one cared very much about what we were doing or thought that it mattered. When we got the minor passed, our male Dean, not a supporter by any means, cautioned us that it wouldn't last forever, to which one of our colleagues appropriately responded "Yes, like Beowolf."

At the same time that I first began teaching Women's Studies, I was working on a Women's Studies text with Alison Jaggar. That text would become *Feminist Frameworks*. Alison, who was then teaching philosophy at the University of Cincinnati, and I, had become friends during the early 1970s and in 1975, both of us attended the National Socialist Feminist Conference in Dayton, Ohio, providing us, and feminists from around the country, with the opportunity to come together. From the beginning of our correspondence, we had been exchanging articles and papers about women's condition. This was the period when women were producing an astounding number of articles at an astounding rate in our efforts to rethink the world but with few avenues for publication. Not only were most prestigious and established journals controlled by men, in addition, there were very few journals that took this kind of material seriously, and in any event, so much was being written that it was impossible to find journal space in which to publish it. This was the period when mimeographed articles began to circulate around the country through a kind of informal underground network. Today those pieces would have found a home easily on the many Web sites that now exist, but before the Internet and before the widespread availability of photocopying, smeared and blurry purple-ink pages were what most of us depended upon. *Feminist Frameworks: Alternative Theories of Relations between Women and Men* was published in 1978 by Prentice Hall and quickly became one of the widely used texts in Women's Studies, ultimately going into three editions.

Unlike other such texts, many of which simply offered an interesting and lively collection of articles, our text offered a structure or a framework that could be used to present and analyze the material in the field. After beginning with an opening section which described "The Need for Women and Men's Liberation," by offering a variety of essays on the kinds of issues that were raised by Women's Studies, Part II, "The Roots of Oppression," presented five alternative theoretical frameworks or viewpoints that could be used to structure approaches to thinking about women's lives. These frameworks used the basic political standpoints to analyze the possible causes of women's oppression by looking at them in terms of the assumptions made by each framework and exploring the implications and consequences that flowed from them. The frameworks we identified were: Conservatism Sexism as Natural Inequality, Liberalism: Sexism as Legal Inequality, Traditional Marxism: Sexism as a Result of the Class System, Radical Feminism: Sexism as Fundamental Inequality, and Socialist Feminism: The Interdependence of Gender and Class.

The third part of the text applied the frameworks to practice by looking at the implications of each of the five theories when they were used to analyze some of the major problem areas that Women's Studies addresses, namely, work, the family, and sexuality. By the time the third edition of the text was published, the frameworks had been expanded so that the section on "Theories of Women's Subordination" now included a section entitled "Through the Lens of Race, Gender, Class and Sexuality: Multiculturalism," and a final section entitled "Women's Subordination World Wide: Global Feminism." Taken together, this way of organizing *Feminist Frameworks* provided a theoretical structure that could be used to model what it would mean if you began with the assumption that the personal is political and used it to analyze alterative approaches to thinking about women's lives. In this way, instead of merely anthologizing interesting pieces, this text provided a way of thinking about feminism that placed these issues in a theoretical framework.

Along with the insight that the personal was political, another early contribution that was extremely important in helping frame and structure Women's Studies as a discipline was an article by Peggy McIntosh titled "Interactive Phases of Curricular Revision." This essay provided a model for rethinking the ways in which the traditional disciplines had been constructed and gave people a concrete example of what a transformed curriculum might look like once it included the experiences and history of the other half of the world's population. According to this model, Phase One offered us a view of womanless history; Phase Two would give us women *in* history; Phase Three would allow us to consider women as a problem, anomaly, or absence in history, Phase Four moves to considering woman *as* history, and Phase Five at long last allows us to think about "history reconstructed, redefined, and transformed to include us all." This article helped many of us understand the inadequacy of the work that had been done before and offered us a vision of what transformation might look like. What would it mean to transform the curriculum? To transform knowledge? What a breathtaking and enormous project—and how wonderful that we could envision it!

Still another of the early and important transformation initiatives, and one which served as a model for many like it, was Wheaton College's *Toward a Balanced Curriculum*. Carried out between 1980 and 1983, The Wheaton Project resulted in a book edited by Bonnier Spanier titled *Toward a Balanced Curriculum: A Sourcebook for Initiating Gender Integration Projects*. The list of contributors included Johnnella Butler, Margo Culley, Beverly Guy-Sheftall, Paul Lauter, Elizabeth Kamarck Minnich, and Betty Schmitz to name a few. As with others like it, this book served as a "how to" manual for many of us engaged in rethinking the curriculum.

During this period and for some time after it, faculty members across the country were involved in creating courses in virtually every field that would show what it meant to create a Women's Studies course in their particular discipline. Many of us gave presentations at conferences and on campuses in which we described these courses and shared the newly created syllabi and bibliographies. They were then independently printed and distributed in newsletters, booklets, books, and other publications and in this way contributed to the proliferation of Women's Studies courses around the country and it's wide distribution. KNOW INC. based in Pittsburgh played a major role in reproducing an enormous number of articles and syllabi and made them available for free.

And then, of course, there were the study groups. There were socialist feminist, radical feminists, and a variety of other reading groups all across the country where people met regularly both to discuss various texts and to plan political actions so that the line between intellectual and political work was virtually seamless. Often it was impossible, and unnecessary, to decide where one ended and the other began. At Paterson State, we met in the Women's Center and in those days the line between Women's Studies as an academic department and the Women's Center and the Women's Collective was very fine—I don't think any of us would have been able to say where one ended and the other began. All of us, the students and the faculty, tenured and untenured, part time and full time, met together to talk about our lives and our situations.

One faculty member, who had mysteriously missed the deadline for having her new Women's Studies course listed on the official printed sheet for spring registration, reproduced her own flyers and posted them on the walls in the women's bathrooms and even in the stalls. Her course was full by the time registration ended!

When we realized that our students had no access to Women's health resources, working together, we persuaded the Student Government Association (SGA) to fund a women's health clinic one day a week where students could obtain a gynecological exam and receive birth control information and supplies if they wanted them. This was very important for our student population because many of them would not otherwise have had easy access to these services. Later, when sexual harassment became a serious issue on our campus, the Women's Collective and Women's Studies faculty played a major role in persuading the administration to create a sexual harassment policy and then played a major role in getting that policy implemented. Establishing a formal lesbian group came later mostly because students (and faculty) were not comfortable going public.

While this begins to give a good sense of what it meant to be part of the creation of Women's Studies, it leaves something out. It still doesn't convey the wild excitement that so many of us felt or the extent of the involvement of students and faculty who saw this enterprise as an extraordinary opportunity to be part of a historic new beginning. For this reason, in trying to capture how best to convey a sense of what it was like to teach Women's Studies at the very beginning, I have settled upon providing a fairly detailed account of my reading list and syllabus for the first Women's Studies course I ever taught. I can think of no better way to capture both the excitement and the intensity of the experience and of our commitment to this new and unprecedented enterprise.

My course, like many such courses around the country, was put together using a combination of paperbacks and a variety of mimeographed articles that I handed out in class or placed on reserve at the library. I required *The Feminine Mystique* by Betty Friedan, *The Subjection of Women* by John Stuart Mill, *The Dialectic of Sex* by Shulamith Firestone, and Frederick Engles' *The Origin of the Family, Private Property and the State*. Some of the readings on reserve were KNOW reprints including Sandra Bem and Daryll Bem's *Homogenizing the American Woman: The Power of a Non-Conscious Ideology*, a piece that still resonates today. There was Margaret Benston's "*Poltical Economy of Women's Liberation*" originally published in *Monthly Review* and then made available as a Warner reprint. There was Evelyn Reed's *Is Biology Women's Destiny* brought out in a Path Finder Press pamphlet, Eli Zaretsky's *Capitalism, The Family and Personal Life* originally published as an essay in *Socialist Review*, and Maxine Williams and Pamela Newman's *Black Women's Liberation*, again put out by Pathfinder Press.

In addition to these required texts, fairly demanding for a state college in the mid-1970s, my syllabus listed a hefty series of supplementary readings for each section. I clearly expected students to refer regularly to *Women in Sexist Society* edited by Vivian Gornick and, of course, *Sisterhood is Powerful* edited by Robin Morgan, with many selections from each book duplicated and handed out in class.

The course began with a section devoted to "Defining the Problem," which was itself broken down into sections that examined "Women's Nature"/Female Sex Roles and Stereotypes; Women at Work, Third World Women, Female Sexuality, and The Masculine Mystique. Ever optimistic about my students' thirst for knowledge and their ability to get by with no sleep, I handed out a selection from Stanley Aronowitz's *False Promises* along with four other articles as required reading for the section on work and placed the just-published book by Harry Braverman, *Labor and Monopoly Capital: The Degradation of Work in the 20th Century*, all 480 pages of it, along with seven other articles on reserve as supplementary reading. So much for the charge that Women's Studies courses were not intellectually demanding.

The section on Female Sexuality included Friedan's interesting critique of Freud, and relevant portions of Simone de Beauvoir's *The Second Sex*, along with other classics of the day: Susan Lydon's "The Politics of the Orgasm," Sidney Abbot's "Sappho was a Right on Woman," and Susan Griffin's "Rape the All-American Crime," which had been published in *Rampart's* Magazine five years earlier. And that's just to name a few.

The section on Men's Lives relied heavily on Jack Litewka's essay "The Socialized Penis," along with Lucy Komisar's "Violence and the Masculine Mystique." They were supplemented by "Measuring Masculinity by the Size of a Paycheck" that I placed on reserve at the library, along with copies of *Playboy Magazine*, John Wayne films, and Lionel Tiger's (give him enough rope and he'll hang himself) "Why Men Need Boy's Night Out."

The section on Third World Women required the classics of the period "Black Women in Revolt," Fran Beal's extremely important article "Double Jeopardy," Eleanor Holmes Norton's "For Sadie and Maud," both of which appeared in *Women in Sexist Society*, and a pamphlet entitled *Black Women's Liberation*, put out by the Socialist Workers Party. In addition to the obvious inclusions from required texts, the supplements for this section included "Gold Flowers Story" in *China Shakes the World* by Jack Belden. Gerda Lerner's *The Black Woman in American*, and Toni Cade's anthology *The Black Woman*, along with two pieces, one by Elizabeth Martinez, that focused on the situation of "La Chicana."

Part Two of the course offered students "A Paradigmatic Account of Interpersonal Oppression." For this, I handed out a photocopy of a portion of Chapter Four from Hegel's Phenomenology ("The Master and the Slave") and supplemented it with selections from Sartre's *Being and Nothingness* as well as from de Beauvoir.

Part Three was called "Alternative Models for Analyzing Women's Oppression" and examined the analysis of the problem and the solutions offered to it by the prevailing theoretical perspectives of the day. There was a section on Liberalism: Reform and Civil Rights that relied heavily on John Stuart Mill's *The Subjection of Women* supplemented by readings from Alice Rossi, Betty Friedan, and Gloria Steinem. This was followed by a section on Orthodox Marxism subtitled "The Primacy of Economic Categories" and included Engles' *Origin of the Family*, the Benston essay, and a somewhat random assortment of additional readings including an essay on "Women: Caste, Class or Oppressed Sex" by Evelyn Reed. Linda Gordon's "The Function of the Family," a piece from Lenin's *The Emancipation of Women* and Peter Laslett's "The World We have Lost." The section on "Radical Feminism: The Primacy of Biological Categories," of course, included Shulamith Firestone's important book *The Dialectic of Sex*, something called The Fourth World Manifesto that I copied from *Female Liberation*, a statement by the NY Radical Feminists, as well as Jill Johnson's *Lesbian Nation* and Roxanne Dunbar's "Female Liberation as the Basis for Social Revolution." The final section of the book and the one that most closely reflected my own political leanings, focused on "Marxist Feminism" and included The Dayton Socialist Feminist Group's "Socialist Feminism: What Does It Mean?," Zaretsky's essay, and an article by myself entitled *"Patriarchy, Private Property and Women's Oppression."* The supplementary readings included Shelia Robotham's *Women's Consciousness, Man's World*, Juliet Mitchell's *Woman's Estate*, an essay entitled "The Power of Women and the Subversion of the Community," and Amy Bridges and Heidi Hartman's "Capitalist Work for Women."

As I look at this ambitious undertaking now, which made demands my students could not possibly have fulfilled, I am struck again by the wonder I felt at this rich and vibrant literature and this extraordinary new premise, which to me at any rate, held the promise of a new beginning. My students were largely working-class, first generation college students from white, ethnic minorities and they were ill-equipped to tackle these readings. Many worked long hours before or after attending class in order to pay their bills. And yet, I do not remember any complaints. For the most part, they, like me, were drunk with the possibility of relevant knowledge and even though I doubt most were able to do more than sample a couple of the supplementary readings, I remember their excitement and gratitude at having been exposed to a little of what was out there and having the option to explore those texts when and if they could. A far cry from the query that so often greets faculty today: "Will it count?" That was a time when everything counted.

And, as I have already pointed out, the line between thought and action was virtually seamless. During the late 1960s, I taught at what was then Hunter College in the Bronx, now Lehman College. I was 23 years old and working on my doctorate and the war in Vietnam framed the context in which many of us thought and studied. In the morning, I taught classes in Political Philosophy where we read

Social Contract theory and debated the legitimacy of the State. We read Thoreau on *Civil Disobedience* and Mill *On Liberty* and in the afternoon we practiced it. Although some philosophers questioned the relationship between thought and action, for me, there was no question. And so, after three years at Hunter when I obtained a tenure-track job at Paterson State, it was natural to bring the same sense of that fluidity with me. I continued to teach Political Philosophy, then Marxism, and continued to be active in antiwar protests. Later, it informed my work in Women's Studies and my work on the campus in general where student empowerment was the order of the day as returning Vietnam Veterans insisted upon having a say in how the college was run.

This was of course, how things should be. If you found a problem, you acted to solve it. When you realized that the lack of childcare on campus made it difficult if not impossible for young women with children to attend school, you pressured the college to create a child-care center. And you found creative ways to do it—such as students bringing their young children with them to classes and even to the Board of Trustees Meetings to get their attention. From the start, the line between thinking and doing was porous in Women's Studies. We didn't want to merely understand the way the world worked, we wanted to change it!

Unsurprisingly, for the first time, some of us began to understand the enormity of our undertaking. The revelations that we came to see challenged the ways in which men and women had come to be defined and made us recognize the way we had come to think about all their lives. One conversation that perhaps typifies the depth of change and challenge this enterprise brought to our lives occurred between the parents of a friend of mine. My friend's mother was in the midst of writing what would become a classic book in feminist analysis, and the conversation was simple and searing. Husband: "If I had known that you had that chapter in you, I would never have married you." Wife: "If I had known that chapter was in me, I might never have married anyone." In that sense, the new perspective that Women's Studies brought encouraged us to rethink everything.

Thinking back to that period, to the early and middle 1970s, I realize how important it was to many of us to link gender issues as seen through the lens of class. At least on the left, an exclusive concern with issues of gender often got you categorized as a "bourgeois feminist" who, to their mind, mistakenly elevated women's issues over class analysis. At that point in time, the theoretical task that occupied many of us was determining where women fit into the substructure/superstructure model that was central to the early writings of Karl Marx. Everybody read and talked about Engels' *Origin of the Family* and legitimizing feminism seemed to require finding a way to place those issues within a narrowly constructed Marxist theoretical framework. The feminists I spent most of my time talking with spent a lot of time arguing about whether women's condition in society could best be understood as a class, caste, or sex or required some other concept. That said, it is important to remember that this discussion took place within the context of a passionate disagreement between radical feminists, lesbian-led, many of whom were separatists, who believed that women had always been oppressed, and socialist feminists who maintained that capitalism was the basis for women's oppression and would have to be eliminated before women could be liberated.

At the beginning, issues of sexuality were probably the topics that received most attention in Women's Studies. But of course, that has changed over time. Over almost 40 years I have been teaching Women's Studies or related courses, there have been times when issues of sexuality were most pressing and easiest to discuss and times when issues of race and ethnicity took center stage. There have been periods when class was foremost in people's mind and times when it disappeared from most peoples' radar entirely.

Initially, the women in my classes were concerned with the politics of sexuality when they focused on sex. They were angry at the double standard that society imposed on men and women and eager to understand how sex had been defined in ways that served men's interests. In addition to focusing on the politics of the orgasm and female sexual alienation, my students began to explore the possibilities of sex between women and many found radical feminism attractive. For the most part, sexuality, like other aspects of life, was understood in the context of "the personal is political" and the women I knew worked hard to understand the ways in which so-called personal choices with regard to sex had social and political consequences that perpetuated or challenged gender roles and social conventions. Their interest was a far cry from the contemporary concern with sex as empowerment that seems to me, in the contemporary period, more about individual empowerment than about empowerment that transforms the social dimension.

What about issues of race during this period? White male liberals and leftists have always been more willing to talk about race and racism than gender and gender issues, while, as with the women's movement a century earlier, white women often found themselves attempting to draw analogies between racial and gender oppression in the hope of legitimating the latter. Naomi Weinstein published an article in *Psychology Today* in 1969 titled "The Woman as Nigger," and a year or two later another writer published a book with that same title.

While some feminists were enraged by what they considered the inappropriate parallels drawn between the treatment of women and that of blacks by American laws, institutions, and culture, during the early period, many or even most white women in the Women's Movement as a whole showed little real interest in analyzing the way women's situations varied based on their race and class or even showed any awareness that these differences existed. Most economic data that was published went no further than providing statistics that broke categories down into male and female but not by race. In addition, women of color did not have the same access, limited though it was, to avenues of publication as did white, male academics. Early women's theory unthinkingly universalized the experience of middle-class white women and in this way contributed to the continued invisibility of women of color, just as men, by universalizing the experiences of men by referring to them as "human beings," had rendered women either invisible or inferior for so long. When Women's Studies talked about "women's" experience, for the most part the women in question were assumed to be white, female academics, and for the most part, this continues to be true today.

On the other hand, from the very start, Women's Studies did include the work and the perspectives of women of color. Robin Morgan's *Sisterhood is Powerful*, which first appeared in 1970, included essays by Frances Beal, Eleanor Holmes Norton, Enriqueta Longauex y Vasquez, and Elizabeth Sutherland while Toni Cade's *The Black Woman*, published that same year, included works by Audre Lorde, Nikki Giovanni, Alice Walker, and many others. They were followed by books like *This Bridge Called My Back*, Barbara Smith's *Home Girls*, *Some of us Are Brave*, as well as important works by bell hooks, Beverly Guy-Sheftall, Johnnella Butler, Patricia Williams, Elizabeth Higgenbotham, and countless other scholar-activists. In 1981, the National Women's Studies Association focused its entire conference on "Women: Responding to Racism." This was a first. There were consciousness-raising groups for white women and women of color, and plenaries and readings by some of the most important women of color writers of the day. Persephone Press (a women's press that has since died) had just published *This Bridge Called My Back: Writings by Radical Women of Color*, edited by Cherrie Moraga and Gloria Anzaldua. The high point of the conference was a debut reading from *Bridge* by the editors, as well as a number of women of color whose worked appeared in the collection (personal correspondence, Lisa Albrecht). Between 1981 and 1983 an important project titled "Black Studies/Women's Studies an Overdue Partnership" codirected by Johnnella Butler and Margo Culley was responsible for generating considerable interest in encouraging this partnership and the resulting bibliographies and presentations were shared at a conference by that name.

Initially, Women's Studies work talked about *integrating* gender into the curriculum or balancing the curriculum, but rapidly this was referred to derisively as "add women and stir" in contrast to the stunning project that most of us embraced, the complete transformation of knowledge. Caroline Heilbrun spoke of "Reinventing Womanhood," one colleague called Women's Studies "a revolution in thought that works," while another announced "what we're all about is the reconstruction of the whole structure of knowledge."

It soon became clear that accomplishing this goal by creating a single "diversity" course to deal with all "diversity" issues was not the way to go. Such courses isolated these areas of concern and thus either allowed students to dismiss them as peripheral to the "real" curriculum or forget their content quickly as they moved through the traditional curriculum. Unless what you teach in Women's Studies and Ethnic Studies is reinforced in other courses, it is easily dismissed as curricular affirmative action—content not believed to be good enough to be included in the general curriculum on it's own merits but permitted to enter the curriculum in order to make people feel better.

In 1986, the State of New Jersey's Department of Higher Education, in collaboration with Women's Studies at Douglass College Rutgers and with a number of two and four year, public and private institutions in the State, created the first statewide curriculum project in the nation, *The New Jersey Project: Integrating the Scholarship on Gender*. During his opening remarks announcing the initiative, Ted Hollander, then New Jersey's Chancellor of Higher Education, acknowledged the stunning importance of the Project when he described it this way: "We undertake this leadership not as a favor to women but as recognition of the urgency to do justice. When gender equality is an integral part of the academic enterprise, then and only then, will we have the moral legitimacy of scholars concerned with truth" (*Newark Star Ledger*, February 20, 1987, 16).

Three years later, when the Project moved to William Paterson University, in Wayne, New Jersey, under my direction, the Project's scope was broadened and it became *The New Jersey Project on Inclusive Scholarship, Curriculum and Teaching*. For me at least, the Project modeled what it meant to take transformation seriously and teach students and faculty to use the lenses of gender, race, class, and sexuality to understand the ways in which a patriarchal worldview and a particular class perspective had been smuggled into the way reality had been constructed.

Women's Studies Today: Disciplining the Discipline

So where are we now? As Women's Studies has made the transition from its beginnings as a liberation movement to an academic discipline, it has undergone the kinds of changes that you would expect. In order to qualify as an academic enterprise, many of us have had to prove ourselves by learning to look, act, and talk like "academics" even as we tried to transform the meaning of what it meant to be one. Those who found the price they had to pay too high often opted for community work. For some, the challenge was how to succeed in academia without compromising their values too much, while others were more than happy to do whatever was necessary to become "legitimate" and move up the career ladder. To this day, some Women's Studies programs and departments are as radical and independent as they were at the start while others seem to have forgotten the meaning of sisterhood.

In my experience, the way people understand the mission and the methodology of Women's Studies depends upon the level of abstraction and the commitment to theory that they bring to their study. Some people are hopelessly bogged down in a superficial understanding of the nature of the work. In sharp contrast, for me, Women's Studies was never about clipping sexist magazine ads—the standard assignment in many introductory Women's Studies classes. While this can be a fine place to begin, if the assignment goes no further, and for many it does not, the discipline can be too easily and mistakenly dismissed. Many Women's Studies courses adopt a superficial approach to gender issues never rising above the amusing but dead-end pages that appear at the back of *Ms.* magazine and that reproduce some of the more egregiously sexist advertisements from popular magazines. But unless Women's Studies students are educated in a way that teaches them about hierarchy and privilege, power and power relations, about wealth and poverty, about oppression and liberation, their education is more than likely to serve as a temporary distraction rather than as a source of lasting empowerment.

From the start, I understood Women's Studies to be a radical undertaking. Indeed, this was part of its charm. It held out to me the possibility of transforming the fundamental assumptions of our society. Anything was possible! I understood Women's Studies to be a radical project in the Marxist sense of radical, as in going to the root cause. In part, this was because my formal academic training was in philosophy. For me the interesting questions were and continue to be meta questions. Questions about how knowledge across the disciplines gets constructed *and whose interests are served by that construction.*

My main gripe with the way so many Women's Studies courses are currently taught is that while some do start with the premise that race, class, and gender are interlocking systems of oppression and hence it is impossible to teach about women without teaching about them as part of a *system*, far too many courses are limited to a focus on women or gender (meaning "women") exclusively. This often leads to the kinds of articles about women found in the *New York Times*. They claim that women are not given their due as workers, wives, or mothers (their standard categories!) and they go no further than turning a few cleaver phrases about women's condition, rarely giving their readers the tools to understand why women always end up with the sticky end of the lollipop. Further, though they are included, women of color and those who identify as LGBTQ are often treated as if they are the only ones who are "different." How else to explain the continuing divide between Women's Studies and Diversity Courses or Multiculturalism? Many such faculty members seem to believe that introductory courses should focus on teaching about gender, by which they mean "women." They are unlikely to consider assigning a diversity text or an anthology that is explicitly multicultural. While it may be appropriate to retitle Women's Studies departments Women and Gender Studies in the interests of inclusiveness, I think that leaving out Women's Studies and in effect subsuming women under gender is indefensible.

What's in a name you might ask? But that would be naive or disingenuous. The recent move away from "Women's Studies" and toward the exclusive use of "Gender Studies," the continuing tendency to bifurcate women and multiculturalism as areas of study, the recent suggestion that it is time to substitute "gender history" instead of doing "women's history," is part of a dangerous and conservatising trend. It encourages us to forget the history that brought us to this point and it obscures the race, class, and gender differences that continue to shape our lives. The reality is that in a world where enormous differences separate people, treating everybody the same will not abolish those differences; it will simply aggravate and perpetuate them while rendering them invisible. But the personal is still political and if we lose sight of this, we lose the perspective that empowers Women's Studies as academics and as activists. We turn from identifying social problems and working to solve them to a world in which individual concerns are the only ones that matter. We forget that knowledge is positional, that the way we see the world reflects the assumptions and interests we bring to our experience so that our experience needs to be deconstructed not embraced uncritically. The critical perspective that defined Women's Studies at the very start, the awareness that the personal is political, was what made the discipline unique and what allowed us to think that we were at the beginning of an extraordinary new project that would ultimately transform knowledge in the interests of us all. That was when sisterhood was powerful.

Note

Attempting to recognize the people and institutions that were part of the enormous project of creating Women's Studies would be impossible. Even as I write this disclaimer, I am embarrassed by the number of important contributors whose names are omitted from my account and I apologize to the many friends and scholars whose names never found their way into this chapter. Hopefully, many of the references and bibliographies attached to this article will remedy some of these omissions. Thanks to Susan Radnor, Barbara Pope, Lisa Albrecht, Naomi Miller, and Judy Gerson for reading this paper. Special thanks to Andrea Mantsios for her loving help and support.

References

Stimpson, C. (1986). *Women's Studies in the United States*. A Report to the Ford Foundation. (See the bibliography attached to this report.)

Braun, R. (1987). "Gender studies called 'Revolution in thought,'" *Newark Star Ledger*, February 20: 16.

Selected Bibliographies

Most if not all of the articles and books mentioned in this essay can be found in one of the following Bibliographies:

Selected Bibliography compiled by Elsa Dixler, included in *Women's Studies in the United States*, Catherine R. Simpson with Nina Kressner Cobb, A Report for the Ford Foundation, 1986.

Guy-Sheftall, Beverly and Susan Heath. (1994). *Women's Studies retrospective*, A Report to the Ford Foundation.

Bibliography in Salper Roberta. (Ed.) (1972). *Female liberation*, New York: Alfred A. Knopf.

Women's Studies Bibliographies, Women's Studies Database, University of Maryland.

Selected Bibliography, Giddings, Paula. (1984). *When and where I enter the impact of black women on race and sex in America*. New York: Bantam Books.

Suggestions for Further Reading included at each section of Jaggar and Rothenberg. (1993). *Feminist frameworks* (3rd ed.), New York: McGraw-Hill.

Moraga Cherrie and Gloria Anzaldua. (Eds.) (1983). *This bridge called My Back*. New York: Kitchen Table Women of Color Press.

In addition, readers are encouraged to refer to the vast array of relevant material that can be found by using the WMST-l File Collection

Readers may also find the following of interest:

Rothenberg, P. (1991). "Critics of attempts to democratize the curriculum are waging a campaign to misrepresent the work of responsible professors." *The Chronicle of Higher Education*, April 10: B1, B3.

———. (2007). "Snatched from the jaws of victory: Feminism then and now." Common Dreams, www.commondreams.org accessed on May 9, 2007.

4

What Took Me So Long?: A Well-Behaved Woman Finds Women's Studies

Jill McLean Taylor

Women's Studies? You teach *what*? What *is* that? You mean you study things like menstruation, pregnancy, and childbirth? Is that appropriate for an university? What about men's studies? Is it fair to have one without the other? Aren't women equal to men now anyway? These and many other questions—some not suitable for print—still greet my response to "What do you teach?" despite the fact that Women's Studies as a discipline has existed since the 1970s as a direct result of activists and academics who were part of the second wave[1] of the feminist movement.

I teach at Simmons College, now a small university with women undergraduates and both women and men as graduate students, where Women's Studies and African American Studies programs began in 1973 and 1972 respectively, the former faculty driven, the latter, student driven, although each was supported by the other group. In the years since they were founded, both have become departments, had name changes that reflected the shift in focus of each, and continued to struggle for resources. A master's degree program in Liberal Studies that was started in the mid-1980s evolved into Gender/Cultural Studies in 1998 with a core of faculty from different disciplines, particularly those with a joint appointment in Women's Studies.

Tracing the history of the department and my own relationship to women's studies, feminism, and a Western feminist movement reveals some of the politics and also some of the ongoing tensions that persist as we address questions of race, ethnicity, social class, sexuality, religion, international women's studies, globalization, ageism, and what appears to be intergenerational conflict in women's studies itself. An example of differences in thinking around about many of the above questions occurred at a Third Wave Feminist Conference at Exeter University that I attended in 2002. A well-known American feminist scholar who was one of the four keynote speakers suggested that there was nothing for feminists to rally around—there was no longer a feminist movement. This was hotly contested by younger women in the audience for whom racism, childcare, access to clean water and adequate food, as well as the right to a healthy sexual and reproductive life were agenda items that must be addressed. All keynote speakers were middle-aged white women, one was transsexual.

As I write about feminism, the feminist movement, and women's studies, it is important to me to identify myself as a Pakeha[2] or European New Zealander for the perhaps obvious reason of reflected glory as New Zealand was the first self-governing country to pass legislation for women's suffrage. On September 19, 1893, the adult women of New Zealand, Maori and Pakeha, won the right to vote, although European women and men voted for the European Members of Parliament three weeks before Maori women and men voted for the four Maori Members (Nolan and Daley, 1993). It was, however, almost 30 years before women could stand for the lower house of Parliament, while the right to be eligible for appointment to the Legislative Council was almost 50 years away.

For the last 11 years, New Zealand's Prime Minister has been a woman, first Jenny Shipley from the National Party, and from 1999 to the present, the Labour Party's Helen Clark, who led the party for six years prior to becoming prime minister and was chair of Foreign Affairs and Defense Select Committee when New Zealand became nuclear free. In 2002, the five most influential posts in New Zealand were held by women. Although many of us have learned over the years, a woman in a leadership position does not necessarily mean change in the ways of doing things, or even in ways of thinking, New Zealand ranks number 5 in the world in terms of gender equity, with the United States at 31 (Hausmann, Tyson, and Saadia, 2007). With Helen Clark as prime minister, there has been some significant legislation affecting women, most notably the Prostitution Law Reform Bill that Helen Clark urged Labour to support (and that passed by one vote), effectively legalizing prostitution. However, critics argue that gender parity is a myth that has caused New Zealand to be complacent, taking the success of a few women at the top as the measure of all women (Hamilton, 2006). An additional point is that New Zealand has had a high gender equity ranking for many years, so that the prime minister cannot take credit for this.

Less obvious reasons for identifying myself have to do with my own lack of awareness of sexism and racism as I was growing up in New Zealand, and, after I left in the late 1960s, the length of time it took me to recognize both good old-fashioned sexism and racism and newer, more subtle varieties. Issues of social class—although not named as such as New Zealand liked to think of itself as a classless society—were more apparent to me as members of my parents' large extended families ranged considerably in their work life and living situations. The "tyranny of distance" is frequently used as an explanation of why New Zealand and New Zealanders may be behind the curve in terms of cultural change coupled with a determination to cling to British traditions, mores, and manners. However, this was not the case for the women's movement in the nineteenth century, nor was it the case in social welfare policies in the twentieth century that were progressive, providing "cradle-to-grave" support. Other explanations of why so many middle-class women seemed unaffected by a feminist movement in the 1960s and 1970s may have to do with the ease of life in New Zealand—a claim not supported by many Maori men and women and many Pakeha women—and the continuing strait-laced attitudes toward sexuality and a fiercely held double standard in terms of sexual behavior. There was also the derision that greeted second-wave feminism, incorrectly associated with bra burning, hairy-legged, unattractive, man-hating women.

Germaine Greer's 1972 visit to New Zealand following the publication of *The Female Eunuch* (1970) challenged all the above stereotypes about feminists and was marked by one charge of obscenity and two for indecent language (later dropped) following her public appearances at Auckland University and the Auckland Town Hall. Amongst other things, Greer addressed the mockery of women's liberation: "Laughter is one of the ways of denouncing the things we are frightened of. Under laughter there is guilt, anger, and hatred: they are not laughing any more in America, it is a genuine political movement, the giggling and snickering in New Zealand is because New Zealand is not used to it yet. The laughter will stop when the country in question grows up" (Russell, 1972). At a lunch on International Women's Day on March 8, 1972, Greer responded to a leaflet describing the lunch as a middle-class affair by claiming that well-educated middle-class women are best equipped to lead a revolution for working-class women, who do not have the freedom to fight, and that being middle class should not be a disqualification. Despite the resistance to feminism, many Pakeha and Maori women, poor, working class, and middle class, were actively working for equal wages, state-run childcare, and reproductive rights, all supported by Greer in her speeches along with the exhortation that New Zealand women should "stop being so damn polite!" (Russell, 1998). My mother sent me clippings with "Daft!" and "She's quite mad!" written above the headlines.

My training and work in New Zealand, England, and Boston at the end of the 1960s and throughout the 1970s was as a physiotherapist (physical therapist), and included teaching natural childbirth classes to pregnant women and, in Boston, their husbands. Some women asked questions about how they might be as mothers, what my life was like, and some spoke wistfully of wanting a life that was different than that of their mothers. My experiences of teaching, plus home visits to patients who were poor and living in Boston in the context of major social changes may have been preparing me to be open to ideas that I read about in the 1970s when my three sons were born and gender roles in my family life were pretty inflexible and obdurate. I was able to work as a part-time physical therapist in what Gloria Steinem named a "jobette" the parameters of which were clear; no disruption of family life; no demands that the other person with a "real job" would be asked to take on any domestic tasks including those related to sick children, and that the money earned would pay for desired goods deemed unnecessary by the head of household. Lest I sound unhappy and embittered, that was not the case. Our friends from both the United Kingdom and Boston were also in traditional roles; women relieved of some aspects of domestic life by a better economic situation than their own parents.

Simultaneously, small intrusions into my formerly well-behaved and nonreflective life started to occur: a discussion with the bank manager when he asked for my husband's signature when I wanted to open my own account; a sideline conversation at our husbands' rugby match with an American woman with three children who was attending law school (and heading law review), who answered "why not?" when someone asked her why; a dinner at Boston University not long after where the difficulty of being a single mother applying and receiving a mortgage was discussed, foreshadowing by 25 years the difficulties recounted by legal scholar Patricia Williams who had the double jeopardy of being a black woman (Williams, 1997).

Although I read about feminists and feminism, I didn't (then) know women who belonged to consciousness-raising groups and I was not aware of the amount of activist work in Boston by black women and white women. Sociologist Winifred Breines describes a complex time in *The Trouble Between Us* (2006), as young women came out of the civil rights movement into the Black Power movement, the New Left, and the student movement they created, the socialist white and black feminist movements in the late 1960s and early 1970s (p. 4). The realities of working across racial differences and, to a lesser degree, class differences made the idealism and universalism of the early years hard to hold onto, but women formed groups, participated in workshops, sit-ins, and protests, organized conferences, and profoundly affected the lives of women and their families in the process. Some of this work was around access to education and opportunities for those who had traditionally been excluded; some was to expand sexual knowledge and openness and to demand reproductive rights including abortion; some was to address violence toward women.

Middle-class white women had an agenda that included access to meaningful work and equity in terms of pay, and they also were critical of a patriarchal family structure that had confined women to narrowly defined roles. The agenda for women of color and poor and working-class white women who had a long history of work was different. Racial solidarity, fighting racism, and strengthening families were of more importance than gender issues for women of color. White women were accused of not recognizing what Peggy McIntosh would call the "unearned privilege of whiteness" and black women resisted overtures by white women to be joined in sisterhood.

Parallel to larger social changes, there were also changes at Simmons in the late 1960s and early 1970s. Black students founded the Simmons Civil Rights Club (SCRC) in 1966 with the purpose of bringing civil rights projects in Boston and at other campuses closer to Simmons. SCRC members also volunteered in the community, and brought speakers to address subjects such as black rights and urban affairs. In 1967 the SCRC became the Black Students Organization (BSO) and on May 5, 1969, black students camped inside the office of the president of Simmons to get a response to 10 demands that they felt had been ignored since 1967. One of these was their demand for a Black Studies program. The following day the president signed the proposal that the demands would be goals to be achieved in the future.[3] In 1971, curriculum committee outlined criteria for a Black Studies concentration that began in 1972 as an interdepartmental concentration. During 1974–1975 the program listed two courses in the catalogue and the name change to Afro-American Studies, which became African American Studies during 1990–1991 to reflect a change in focus. In 1997 African American Studies became a department, and in 2001 a further name change to Africana Studies signaled that interests of the department had expanded to include black peoples in Africa, the Caribbean, South America, Canada, and Europe. Throughout the years, the BSO has maintained a strong and active presence on campus.

Women's Studies at Simmons began in response to a critique of traditional disciplinary practices, and in support of the mission of the college. It started with two courses in 1973, became a program in 1975 when the catalogue described Women's Studies as having been designed "both to identify and to encourage the development of courses that deal with women's experiences. Through an understanding of the ideas and variety of women's activities viewed across time, across culture, and from the perspective of various disciplines, the program hopes to help women appreciate their potential as well as their heritage" (Simmons Catalogue, 1975/1976, 146). In 1981 the program offered a 36-hour independent concentration (major) as well as a 36-hour dual concentration with course requirements that remained basically the same with some additions and changes in course levels until 2005–2006. Following African American Studies by two years, Women's Studies became a department in 1999, and after a great deal of discussion that revealed faculty members' different views, changed its name to Women's and Gender Studies in 2007, better reflecting our curriculum and also student interest. Some wanted to add sexuality but our course offerings do not include enough to make this appropriate. The decision to keep women in the title rather than moving to gender studies was for women's studies to be a priority and more prominent than gender, and to make sure that feminist scholarship is not depoliticized by moving too far away from the study of women. Historian Joan Scott has noted that the use of the term gender fails to name the aggrieved—women.

As one of the first programs in the New England area, Women's and Gender Studies at Simmons has continued to be inter and multidisciplinary through various changes, allaying fears voiced by faculty as we moved toward department status that we would not only give up interdisciplinarity but that we would also become institutionalized, that we would lose any epistemological advantage from the margins. Practical concerns about the ability to "borrow" faculty were also raised, but countered by the advantage of being able to have a core of faculty with formal appointments who would be committed to Women's Studies and thus have the ability to offer a planned and stable curriculum into the future. Using memorandum of understanding between faculty members' two departments has been useful, but it is more often the "home" department that prevails around course decisions and teaching load. We have been affected by the success of various faculty members with joint appointments who have become senior administrators and chairs of their original (hiring) department at the college, and although another faculty member and I came up for tenure as joint appointments, and a junior faculty member was hired as a joint appointment in Women's Studies and Africana Studies, we have not yet received another faculty line. The recognition that we are doing more with less *and* not complaining—except to one another—is familiar and distressing, although, as with other institutions, we are by no means the only department wanting additional faculty.

While changes were taking place in colleges and universities in Boston at the end of the 1960s and early 1970s, activist groups such as Bread and Roses and the Cambridge Women's Center also included education for women who had traditionally been denied access to postsecondary education, or to finishing secondary education. Bread and Roses had organized a feminist takeover of a Harvard Building at 888 Memorial Drive on International Women's Day in 1971 that lasted for 10 days and pushed Harvard to provide space for a center that would include not only Harvard women, but also women from the community. In 1972, a year after the occupation, the Cambridge Women's Center had been set up without help from Harvard. Bread and Roses founded the Women's School at the Center, and groups and individuals through Women's Inc. and other community-based organizations set up educational programs for women. The health collective that produced *Our Bodies, Ourselves* began in a class offered by Bread and Roses before the school was founded. In keeping with their belief that children were the responsibility of a community, free childcare was available for women taking courses (Breines, 2006, 103–105).

Similar to the education being offered by community-based groups, the College for Public and Community Service (CPCS) at the University of Massachusetts at Boston was founded in 1972. The CPCS, like the Women's School in Cambridge, was radical in its approach to educating nontraditional students. Most were poor or working class, a high proportion was from diverse racial and ethnic groups, the average

age was 35, and many had children and were employed full-time. The college is competency-based in that students receive credit when they demonstrate competencies such as analyzing arguments, basic social criticism, taking a stand, and advocacy speaking, and the curriculum is grounded in a critical perspective, linking "where possible theory and practice and making instructional activities relevant to students' lives as workers and citizens" (Steinitz and Kanter, 1991, 140).

I found my way to CPCS as a student in the fall of 1980, thanks to another childbirth instructor, and I trace to there the beginning of both formal and informal "consciousness raising" and education that caused me to question and challenge so many of the things that I had accepted as the way things were because of gender differences. In fact, to begin with, I struggled with the idea that gender and race were socially constructed—now, almost 30 years later, that struggle seems to be from another life. I see now that courses were feminist in how they were taught, and the perspective from which they were taught. An example was the advanced concentration in Women's Studies cotaught by an historian and a philosopher to cover the time periods 1810–1860, and 1930–1980 and to address the competencies Explaining Historical Change, and Evaluation of Historical Change. I have kept my typed paper on "Childbirth Practices in America: The Changes Before and During the1930s to the 1980s," as it not only represents taking myself seriously, but being taken seriously, an integral part of a feminist classroom.

During my first term at CPCS—after a suburban dinner party where, to my dismay, the conversation seemed to support Shockley's argument about race and intelligence, and when some of us tried to argue against this, we seemed to be speaking a different language—an article appeared in the Sunday, October 5, *Boston Globe Magazine*. It was written by journalist Christina Robb (recent author of *This Changes Everything*), who described the work of New Zealander, Susan Moller Okin, a political philosopher at Brandeis University, and that of an American psychologist, Carol Gilligan at Harvard. A radical notion that because things were a certain way in my family in New Zealand when I was growing up they did not necessarily have to be the same in our family in Boston took hold in conjunction with the recognition that Gilligan's 1977 Harvard Educational Review article and the work that became *In a Different Voice* made sense to me and to thousands of other women. My sons were young, my Northern Irish husband preoccupied with his career; I had gathered courage and telephoned Susan Okin to talk about growing up living close to one another in New Zealand; I was reading and discussing Marxist and feminist theory; I had read and been with my classmates to hear Jean Baker Miller talk about her 1976 book, *Toward a New Psychology of Women*: Anything seemed possible.

As many other middle- and upper-middle-class white women of about the same age were to discover, what *seemed* possible and what *was* possible were two different things. We could have saved ourselves a lot of time had we listened to black and Latina women talking about their experiences of how power and privilege function in relation to the intersection of gender, race, social class, sexuality, and nation. And we could have examined more closely and addressed our assumptions about race, ethnicity, and class, and what it meant to work with others from different backgrounds.

By the end of the 1970s and early 1980s, black and white activist women, only some of whom identified themselves as feminists, were working together in Boston and in other parts of the country (Breines, 2006, 104). Much of this cooperation and cross-racial learning was around safety for women in response to violence, and in particular a number of rapes of white women in the winter of 1978–1979 that were followed by the murder of twelve black women and one white woman in Roxbury between January and May of 1979. The multi-racial Coalition for Women's Safety was formed in reaction to the violence and the racist and sexist response of the police and the media. The Coalition is cited as an example of white women, primarily from community-based groups, working as allies with women of color. Given the deeply divisive school desegregation busing in 1974 and 1975, it was extraordinary that people could work across their differences. This was articulated in a letter on June 26, 1979, by Barbara Smith, a black radical activist who was a leader in the coalition and other groups:

> We need to talk about what it means for Black and other women of color to be working with white women, for white women to be working with women of color, for women who identify as feminists to be working with women who do not identify themselves as feminist and for Lesbians and heterosexual women to be working together. A discussion of class difference[s] may also be useful…. Since these differences are definitely there it only makes sense to speak to them. Not for the purpose of divisiveness, but for the purpose of understanding and greater closeness. I feel proud of what we have done so far and how well we have dealt with each other under so much pressure. (cited in Breines, 2006, 169)

Feminists may have been able to find ways to work across differences, but that was not my experience as an individual in daily contact with another woman. Sharon, the day care provider for my oldest son, in the mid-1970s, was white and living in the city with her husband and three children, who were bused to school as part of the school desegregation plan. Like other parents, the safety of her children was her chief concern, especially after one had a minor injury when the bus was forced to stop suddenly. She was angry and scared and it became difficult for us to speak with one another about what was happening, especially as by that time I was living in a suburb close to her, but outside the city, and so my children would not be affected by busing. We avoided speaking about race and what was happening: we seemed to have no language to understand our different social-class positions.

In 1982 I graduated from CPCS with a Bachelor of Arts in the Management of Human Services, identified myself as a feminist, and in the fall began a Master's Degree Program at the Harvard Graduate School of Education. I had been to hear Carol Gilligan speak and requested (if accepted) that I was an advisee. I also was able to take half my classes in the Maternal and Child Health Department at the Harvard School of Public Health to continue learning from that perspective about sexuality, reproduction, and motherhood, particularly in adolescence. *In a Different Voice* had just been published; Lawrence Kohlberg and Carol Gilligan and their students were engaged in conversations about moral development, about adolescent identity formation, about different worldviews that were related to gender and morality. My long-suffering friends and their children began to tire of Heinz and I looked forward to beginning the doctoral program in Human Development and Psychology in the fall of 1984, the same time as Carol was *Ms.* magazine's Women of the Year and came up for tenure and promotion.

Meanwhile, the acclaim for *In a Different Voice* was not universal. Some academic feminists seemed to misread and oversimplify her work, others vehemently disagreed with her, others pointed to the lack of diversity in the participants in the studies, others criticized the research methods, and others complained that the ideas were not new (See *Signs*, 1986). At the same time, book sales flourished, discussions and citations of Gilligan's work continued to grow, as did support for her research with girls in independent school settings, which later expanded to adolescent boys and to boys and girls in urban Boys and Girls Clubs. A group of HGSE (Harvard Graduate School of Education) students worked with Carol at the Center for the Study of Gender, Education, and Human Development, later changed to the Project on Women's Psychology and Girls' Development. Although I was not there for the Emma Willard Study, I was involved with others that included the Laurel School in Cleveland, Ohio, the Middlesex School in Concord, Massachusetts, and the Boys and Girls Clubs Study in Boston.

I was also doing doctoral research with a small group of adolescent mothers in an Adolescent Parenting Program at Cambridge Rindge and Latin High School and working to secure a grant to study adolescent decision making and identity formation with a group of "at risk" students. This, the Understanding Adolescence Study, allowed us to listen to girls considered to be at risk for early pregnancy, and/or dropping out of school and to include what we learned in understanding girls' development and women's psychology. I remember clearly the challenges involved in trying to get funding for "at risk" girls, especially compared to the willingness of parents and boards of trustees to come up with funding for research at independent schools. It was, after all, the Reagan mid-1980s and only six or seven years after Joseph Adelson's remark that there was not a chapter on girls in the 1980 *Handbook of Adolescent Psychology*, as a leading scholar had concluded that there was not enough good material to warrant a separate chapter. The politics of bringing girls' voices into considerations of psychological theory was heightened when the girls were poor, black, Latina, Portuguese, from single-mother families—the underlying question seemed to be what could be learned from them when they were clearly "deviant." Fortunately, Wendy Puriefoy, then at the Boston Foundation, thought there was a great deal to be learned, and urged us to connect the program to Boston by working with teachers, social workers, and guidance counselors from two Boston Public Middle Schools.

The late 1980s into the mid-1990s was a period when questions of difference were central. Our research group prepared to interpret the data we had collected over the first two years of the study with girls who were from racial, ethnic, and social class backgrounds that were different in many ways from the predominately white researchers. A crucial element of data analysis was setting up an interpretive community as our focus on race, ethnicity, and social class underscored the importance of "who is listening" as well as "who is speaking." Before beginning the third and final year of interviews, we invited Michelle Fine and Janie Ward to work with us to reexamine our research questions and explore ways of refining or redirecting our inquiry. Janie noted that in my eighth grade interview with Anita, a black girl I had interviewed in both the seventh and eighth grade, Anita had attempted on three occasions to introduce the topic of race in response to a question about something that made her feel bad about herself in school, and that I had not picked up on it then and neither had the interpretive community. Michelle followed Janie's observation with the suggestion that we interrupt the cultural taboo on speaking directly about race into our interview with girls, which led to asking ourselves how deeply we were willing to listen and speak about race and ethnicity ourselves (Taylor, Gilligan, and Sullivan, 1995).

Concurrently, as suggested by the Boston Foundation, we were beginning a series of retreats with women from the Boston Public School in a project called Women Teaching Girls. To prepare for this we invited white women and women of color to join in a weekend retreat exploring relationships among women and also between women and girls across politically and economically significant racial and cultural differences. This first retreat expanded into six Women and Race retreats over a two-year period and involved eleven women: five black, five white, and one Latina. The retreats took as their starting point a question that was made clear from our attention to girls: will women—will we—perpetuate past divisions among women into the future, including racial and class divisions that have been so psychologically and politically divisive and painful? As we listened to girls in the Understanding Adolescence Study we found our research expanded into political inquiry: can women act in concert to end a racist and sexist society? It turned out to be much harder than some of us had initially thought to understand each other in a group of eleven, never mind ending a racist and sexist society.

While working as project director for the Understanding Adolescence Study, I finished my doctoral work with my research on adolescent mothers, graduated in 1989, and a year later, went to Simmons College as a year-long replacement for Janie Ward in the Department of Education and Human Services. As Janie returned, a position opened in the department working with grants with the Boston Public Schools and I stayed at Simmons. I was working with the study, the middle school project, Women Teaching Girls, to which we immediately added Girls Teaching Women as the second part of the title, and with the Women and Race Retreats. The chair of Education, Kay Dunn, invited me to the Women's Studies reading group that was started in 1990, and it was through this group that I was asked in

1994 if I would teach Women's Studies 100, the introductory course for the major, which I have done almost every year since.

For that first year, I had a large photocopied packet that contained articles of what were considered foundational texts for women studies (this was prior to changes in copyright laws). The packet of well-worn originals contained essays and chapters by Phyllis Chesler, Cynthia Enloe, Anne Fausto Sterling, Carol Gilligan, bell hooks, June Jordan, Judith Lorber, Audre Lorde, Adrienne Rich, Carol Sheffield, Gloria Steinem, Iris Young, and others. Many of these articles are in a number of anthologies that I have used with one or two other books, such as Leslie Feinberg's *Stone Butch Blues* (1993). Recent class texts include *The Politics of Women's Bodies* (Rose Weitz, Editor, 2nd edition 2003); *This Bridge We Call Home* (Gloria Anzaldua and Analouise Keating, Editors, 2002); *Colonize This!* (Daisy Hernadez and Bushra Rehman, 2002), and reserve has allowed students to download readings.

Current students, some of whom identify themselves as third-wave feminists, are interested in sexuality in general and transgender and intersex issues in particular. In 2005–2006, curricular changes included the return of Introduction to Lesbian, Gay, Bisexual and Transgender Studies, which had only been taught once before. Women in Literature or Women and Work, which became double-listed in both English and Women's Studies and Economics and Women's Studies, along with the two introductory courses described above can now serve as the entry-level course to the major or minor in Women's Studies. In the future I believe that students will demand further courses on gender and sexuality, Queer Studies, and ethnic studies. Our Women, Culture, Nation course that is a required mid-level course is also required for International Relations students. Currently, we have an upper-level course that graduate students also take, but have to rely on a sociology course, Body Politics, for an intermediate level course. The Biology of Women course is a sought-after elective, and has been one of the department's successes to have a student who was a double major in Biology and Women's Studies win the most prestigious academic prize at the College, putting to rest the questions of intellectual rigor that dogged women's studies in the early days. Another student double majoring in Philosophy and Women's Studies has also received this award.

I became tenure track in 1996, and was granted tenure as the first joint appointment in the college in 2000. Since then I have served as interim chair, interim program director for Gender/Cultural Studies, and chair of Women's Studies for the last four years and have taught other Women's Studies and Gender/Cultural Studies courses including a course for Social Justice Minor. At the same time, I have been involved in urban education as coordinator of GEAR UP for seven years and through that have been able to work closely with Latina adolescents at a high school, as well as black adolescents and Latinos most of whom are labeled "at risk" or "high risk" even though they have such enormous potential and resilience.

At the risk of sounding as though I am writing a college application essay, or preparing an interview for the Miss America pageant, Women's Studies, introduced by looking closely at women's lives through something I knew—childbirth education and practices— and then expanding to include human development and psychology and other disciplines, has given me perspective and ways of thinking and understanding. Listening to girls talk about their lives required recognizing my resistance to taking myself seriously as a woman; working with women across race and class and sexual and cultural differences required seeing myself as having the unearned privilege of being white in a society where race still matters. And I think of teaching and identifying myself as a feminist as a political act, one that requires transgressing by speaking out and listening to young women in a time when freedom and liberation from poverty and violence and racism and sexism for many women and girls is still an ideal and not a reality.

Notes

1. Although the "wave" concept of the women's movement continues to be used, some feminists argue that this metaphor suggests that between the waves there is little happening.
2. Pakeha was the name given by Maori to whites. Originally thought to mean white person, the translation into Maori is closer to "go away," or in the New Zealand vernacular, "bugger off."
3. The information on the BSO demands comes from archival material including the Simmons News on May 8, 1969; personal communication with Professor Emeritus, Mark Solomon, and the African Studies Self-Review Report, 2007.

References

Anderson, M., and H.C. Patricia. (Eds). (2006). *Race, class and gender* (6th ed.). San Francisco: Wadsworth.

Anzaldua, G., and K. Analouise. (Eds.). (2002). *This bridge we call home: Radical visions for transformation*. New York and London: Routledge.

Baillet, B. (Ed.). (2008). *Women, culture and society women: A reader* (5th ed.). Iowa: Kendall/Hunt.

Breines, W. (2006). *The trouble between us: An uneasy history of white and black women in the feminist movement*. New York: Oxford University Press.

Feinberg, L. (1993). *Stone butch blues*. New York: Firebrand Books.

Gilligan, C. (1982). *In a different voice*. Cambridge: Harvard University Press.

Greer, G. (1970). *The female eunuch*. New York: Farrar, Straus and Giroux.

Hamilton, J. (2006). "International women's day a reality check for New Zealand." http://www.wunrn.com/news/2006/13_05_06/031006_new_zealand_htm accessed on January 23, 2008.

Hausmann, R., L.D. Tyson, and Z. Saadia. (2007). "The Global Gender Gap Report. World Economic Report." http://www.weforum.org/en/inititaives/gcp/Gender%20Gap/index.htm accessed on January 23, 2008.

Hernadez, D., and R. Bushra. (Eds.). (2002). *Colonize this! Young women of color on today's feminism*. New York: Seal Press.

Nolan, M. and C. Daley. 1993. "International feminist perspectives on suffrage: An introduction." In C. Daley and M. Nolan (Eds.), *Suffrage and Beyond: International Feminist Perspectives*. New York: New York University Press, 1–22.

Russell, M. (1972). "Greer in action-A diary." *Thursday Magazine*, Auckland, New Zealand. March 16.

———. (1998). "Wish lists: Beyond suffrage." In *Book of the century*. Auckland, New Zealand, *New Zealand Herald*, 91–103.

Steinitz, V. and S. Kanter. (1991). "Becoming outspoken: Beyond connected education." *Women's Studies Quarterly, Women. Girls, and the Culture of Education*, 19(1 and 2): 138–153.

Taylor, J., C. Gilligan, and A. Sullivan. (1995). *Between voice and silence: Women and girls, race and relationship*. Cambridge: Harvard University Press.

Weitz, R. (Ed.). (2003). *The politics of women's bodies* (2nd ed.). New York and London: Oxford University Press.

Williams, P. (1997). "Of Race and Risk." *The Nation*. December 29.

5

Women's Studies: A View from the Margins

Beverly Guy-Sheftall

There is a growing body of work that assesses the field of Women's Studies, now almost 40 years old, if we use as a starting point 1969, the launching of the first program at San Diego State University. I'm referring to *Women's Studies for the Future: Foundations, Interrogations, Politics; Women's Studies On Its Own: A Next Wave Reader in Institutional Change* (Wiegman, 2002); *Troubling Women's Studies: Pasts, Present, and Possibilities* (Braithwaite and Heald, 2005); and my own consultant's report, *The Status of Women's Studies, 1969–1992*, which the Ford Foundation commissioned and was published in 1995 as *Women's Studies: A Retrospective* (Guy-Sheftall, 1995b).[1] Among the retrospectives I probed for this essay, Elizabeth Kennedy and Agatha Beins' (2005) anthology, *Women's Studies for the Future,* raised the most compelling questions for me: What is the subject of Women's Studies? How does Women's Studies negotiate the politics of alliance and the politics of difference? How can Women's Studies fulfill the promise of interdisciplinarity? What is the continuing place of activism in women's studies? And finally, how has feminist pedagogy responded to changing social conditions?

Initially, my desire was to organize this retrospective around how I might answer these questions from my particular location as the founding director of the only women's research center at a historically Black college with both a women' studies major and a women's center. The Spelman College Women's Research and Resource Center, founded in 1981, is also the only women's studies institute in the academy with an LGBT (lesbian/gay/bi-sexual/transgender) program, an archival program, a student activist leadership project, linkages with universities and NGOs (nongovernmental organizations) throughout Africa and the African diaspora, a digital media project, a global women's health initiative, and a commitment to becoming a major repository for materials on Black women in higher education and the contemporary Black feminist movement. The Women's Center, which celebrated its twenty-fifth anniversary in October 2006, will also be the first academic program at the College with its own separate endowment.

Though I was tempted to answer the questions posed by the editors of *Women's Studies for the Future*, I decided instead to write a more personal narrative. I wanted to reflect more deeply upon why I was drawn to women's studies, why I'm still passionate about the field, despite all its flaws and challenges, and why I think it's important for young women and men to read about the joys of feminism and the life we chose as women's studies scholar/activists.

In many ways the genealogy of my own professional life provides a glimpse into the transformative impact of women's studies *and* feminist politics on those of us who came of age in the sixties. I can say without equivocation that my involvement in the development of the field changed my life. Unlike many feminist academics, however, I decided to do my radical political work not at a majority institution but at a historically Black college for women whose evolving mission I wanted to help shape. My teaching career began, however, in 1969 at Alabama State University, a black public university, the same year that the first women's studies program was initiated on a very different campus. I began teaching English at Spelman College, my alma mater, two years later. It was here where I learned, without the benefit of mentors, how to be an effective force for change in a context where my own ideological, political commitments around feminism were not necessarily shared, especially among my students initially.

Five projects in which I became engaged as a junior faculty at Spelman had a significant impact on my continuing journey: co-editing with a colleague the first anthology on Black women's literature, *Sturdy Black Bridges* (1979); teaching the first women's studies courses within our Eurocentric English department in the early 1970s, long before the institutionalization of women's studies in the academy—Images of Women in Literature, Images of Women in the Media, and Black Women Writers; participating in the lock-up of the Board of Trustees because of their appointment of another male president in 1976; and founding the Women's Research and Resource

Center in 1981, while I was untenured and a doctoral student; and establishing the first women's studies program at a historically Black college which started with a minor and became a major in 1996.

In 1974, three years after I joined the faculty at Spelman College as a member of the English Department, I met Toni Cade Bambara who was to have the most profound impact on my career as a feminist scholar/activist. Because of the significance of her anthology, *The Black Woman* (1970), and her importance as a writer, educator, feminist critic, and community worker, I decided to interview Toni (in her Atlanta home) for *Sturdy Black Bridges*, my first publication, which was co-edited with Roseann P. Bell, my English department colleague. Though I would never have been able to anticipate Toni's impact, this first meeting was pivotal in my evolving feminist consciousness. When I inquired about the possibility of Black and other third world women forming alliances around the eradication of both race and gender oppression, Toni revealed her involvement in what we would now call a global women's movement. A year later (1975), she would travel to North Vietnam having been invited by their Women's Union with a delegation called the North American Academic Marxist-Leninist Anti-Imperialist Feminist Women. She believed that feminist movements should not be narrowly focused on the evils of patriarchy but rather on eliminating all oppressions, including that experienced by the colonized third world and poor people in the United States.

She also bemoaned a missed moment in the early 1960s when what she called a national Black women's movement could have been multicultural and multiethnic if African American women had joined Puerto Rican women and Chicanas in their struggles for liberation. When I asked her whether it was a dilemma for her to be both a feminist and a warrior in the race struggle, which for some Black women and men was an oxymoron, she asserted unequivocally, "I don't find any basic contradiction or any tension between being a feminist, being a pan-Africanist, being a black nationalist, being an internationalist, being a socialist, and being a woman in North America."[2] This was precisely the message many of us young Black feminists needed as we found ourselves increasingly under suspicion with respect to our race loyalty.

My decision two years later to enroll in a particular doctoral program at Emory University, the Institute for Liberal Arts (ILA), was connected to the kind of scholar/ activist I wanted to become. I chose Emory's American Studies program rather than the Ph.D. in English because I could pursue my interests in both African American Studies and Women's Studies and because I knew the ILA to be an oppositional space within American higher education—a place where interdisciplinarity was at the core of the program, where transgressive intellectual work would be supported, and where young scholars like myself, especially young Black women scholars, would never have to apologize for our interests in race and gender, or struggle for our voices to be heard. I also wanted to involve myself in the kind of intellectual and political work that Toni Cade Bambara's life represented.

As I had hoped, my evolving love affair with transgressive knowledges and oppositional pedagogies was nurtured at Emory among a small group of faculty who were committed to alternative and boundary crossing ways of doing serious academic work. Unlike many students of color, I did not leave Emory scarred, battle-weary or cynical about the academy. I left there equipped with knowledge from various disciplines (literature, history, sociology, art history) and deeply committed to interdisciplinarity, particularly the fields of African American Studies and Women's Studies; a clear vision about how I wanted to spend the remainder of my academic life; and with passion about my participation in the development of women's studies nationally.

I also left Emory with a serious commitment to continue my scholarly work on Black women that began with a U.S. focus but now encompasses the experiences of women of African descent globally. Shortly after Emory established the first doctoral program in Women's Studies, I was invited to teach as an adjunct professor during which I developed two courses—"African American Women's History" and "Black Feminist Thought"—that resulted in the publication of *Words of Fire: An Anthology of African American Feminist Thought* (1995b). I had also become frustrated about the failure of women's studies to take seriously the feminist theorizing in which African American women had been engaged since the nineteenth century. I was also tired of refuting the case within African American communities that Black women were uninterested in feminism or simply responding to the racism of white women when they engaged in gender discourse. Eventually, I began to teach a new course at Emory, "Global Black Feminisms," which focused on feminist thought and activism among women of African descent in Africa, the Caribbean, Brazil, Canada, and England.

My evolving interest in Black feminist studies and its global dimensions led to my moving away from teaching graduate courses at Emory that focused narrowly on the United States. While the field of Women's

Studies was also becoming more global, women in Africa and the African diaspora were still largely invisible as scholars or subjects for classroom study. Despite the pervasiveness of Western feminist theory in women's studies classrooms, I was becoming increasingly aware of the growing body of creative, scholarly, and political work by African and African-descended women that could be placed under the rubric of African feminism. Any discussion of African feminism is likely to make reference to the foundational work of scholars such as Molara Ogundipe-Leslie, a Yoruba writer, critic, intellectual and scholar who has been writing about the issue of gender politics and social transformation within the African context for over three decades. One of the most compelling treatises in her collection of essays, *Re-Creating Ourselves: African Women and Critical Transformations* (1994), is "Stiwanism: Feminism in an African Context," which I would compare with the Combahee River Collective manifesto which captures the essence of African American feminism, namely the idea that black women experience multiple jeopardizes, including racial, class, and gender oppression.[3]

Ogundipe-Leslie was a founding member of Women in Nigeria (WIN), a feminist women's research and activist group, as well as of another African feminist organization, the Association of African Women for Research and Development (AAWORD), located in Dakar, Senegal. One of the leading African woman intellectuals, whose work is largely unknown in mainstream women's studies circles, she argues that gender asymmetry was an integral part of many precolonial societies, even though African women's position relative to men certainly deteriorated under colonialism.

She also argues, as do African American feminists, that there are also traditions of resistance and activism among African women going back to precolonial times. In other words, there were indigenous "feminisms" prior to contact with Europeans, just as there were indigenous modes of rebellion and resistance throughout the period of colonial domination. What this means is that the struggle for women's rights is not the result of contamination by the West or simple imitation by African women of Euro-American values.

In addition, she reminds us that in most cultures there have always been indigenous manifestations of "feminism" that take various forms, albeit very different forms, in particular cultural contexts. An important task for African scholars has therefore been to identify, excavate, and analyze these indigenous forms of feminist resistance. Bisi Adeleye-Fayemi's essay, "Creating and Sustaining Feminist Space in Africa" is one of the most compelling review essays for Western scholars interested in revisioning and teaching about the meanings of feminism in African cultural contexts.[4]

An outpouring of Black feminist intellectual production over the past two decades in Africa, the United States, Canada, the Caribbean, Latin America, and Black Britain, in particular, underscores the importance of taking seriously the voices of women of African descent with respect to a range of social justice issues at the intersection of race, class, ethnicity, and gender. A comprehensive examination of African feminisms, both the scholarship and activism, is certainly long overdue. Such a sustained project would help to contest the hegemony of Western, white feminist analytical frameworks and broaden our knowledge about a region of the world that continues to be misunderstood, devalued, and, in some cases, romanticized, especially with respect to gender constructs. I do believe that the scholarship by and about women of color around the globe has reinvigorated women's studies in such a way that it is not as narrow or mainly a manifestation of one group of women's experiences.

While Women's Studies has continued to evolve in profound ways since its inception in the 1970s, my greatest disappointment is captured, however, in the opening paragraph of a recent essay in *Meridians* that echoes the sentiments of many feminist women of color who continue to struggle with the field: "White feminists experience extreme levels of discomfort when an antiracist perspective is introduced within feminist sites of engagement, including academic meetings, professional organizations, conferences, and classrooms."[5] Over the past several decades, to be sure, women's studies as an intellectual and political project has gained visibility and importance around the world. Feminists everywhere have variously and successfully transformed lives, communities, and institutions. Nevertheless, questions of social and economic justice, identity and self-determination, psychic and social decolonization, and solidarity and alliance-building across class, race, sexual and national borders, remain at the heart of feminist work and is our biggest challenge going forward.

In order to continue to participate in the transformation of women's studies, the Women's Center at Spelman has engaged in broader feminist networks and hosted an international conference on Women, Girls and HIV/AIDS in Africa and the African Diaspora which attracted delegations from South Africa, Senegal, and Brazil; this was the first gathering at a Black college to discuss both the racial and gendered dimensions of the pandemic. Our evolving work on African feminisms (Associate Director M. Bahati Kuumba guest-edited a special issue of *Agenda*, a South African feminist journal) has resulted in our establishing collaborations with women on the continent. In January 2005, we took a delegation of faculty and students to South Africa to attend the "Writing African Women: Poetics and Politics of African Gender Research" conference at the University of the Western Cape. In October 2005, a group of feminist scholar-activists from Brazil, the Caribbean, West and southern Africa came to the Center to formalize our network.

One result of this historic gathering was the idea to host a faculty seminar on "New Research and Discourse on Gender and Sexuality in the African Diaspora," which took place on the Spelman campus in May 2007. The readings represented the perspectives of Black feminist intellectuals and others located differentially throughout Africa, the Caribbean, and the Americas. We identified and explored the emerging themes and directions of this growing body of scholarship from an interdisciplinary, intersectional, transnational, (her) historical, and critical feminist perspective. The seminar's intellectual content represented the cutting edge of critical Black feminist theory and analysis as applied to a range of recent events and areas of inquiry, that included the aftermath of Hurricane Katrina, wars and militarization, human rights, neoliberal globalization, and expressions of alternative sexualities in particular cultural contexts throughout Africa and the African diaspora. Interdisciplinary readings such as Aaronette White's "All the Men Are Fighting for Freedom, All the Women are Mourning Their Men, but Some of Us Carried Guns: A Race-Gendered Analysis of Fanon's Psychological Perspectives on War," was one compelling example of the impact of *But Some of Us Are Brave* as well as new directions in Black feminist studies.[6]

On a personal level, I have collaborated with other women of color in order to deepen my own understanding of the global realities of women as well as become more engaged in networks away from the Spelman campus that have been deeply satisfying and intellectually engaging. In this regard, Professor Chandra Talpade Mohanty (Women's Studies, Syracuse University) and I have been teaching summer faculty development seminars at New York University's Faculty Resource Network to a remarkably diverse group of faculty from the United States and the Americas. This summer we will teach "Women's Studies in a Global World: Pedagogies of Transformation," that will provide historical, critical, and epistemological perspectives on the development of women's studies, feminist practice, and its connections to multiple social movements in Asia, Africa, Latin America, the Caribbean, Europe, and the United States. Mohanty and I also co-teach a women's studies faculty seminar at Cornell University under the auspices of a new mentoring program for faculty of color, The Future of Minority Studies, which is funded by the Mellon Foundation.

Notwithstanding these exciting off-campus feminist collaborations, another disappointment for me surrounds the issue of the institutionalization of women's studies in the academy. Despite the fact that there are over six hundred undergraduate women's studies programs in U.S. colleges and universities and a growing number of doctoral programs, women's studies is still institutionally fragile in most places, in the sense that most women's studies programs do not have departmental status and are therefore without their own faculty lines; they also have inadequate budgets and very little control over their curriculum because they depend on departmental courses or joint appointments. There has also been in recent years a tremendous backlash and tightening of resources so that though women's studies is visible in the academy and has had a tremendous impact, it is not as institutionally strong in most traditional departments or disciplines.

After forty years of involvement with Women's Studies, which actually began when I was writing a thesis on Faulkner's treatment of women in his major novels in 1967 at Atlanta University, I am as committed to this interdisciplinary field as I was when I enrolled in my first women's studies course as a doctoral student at Emory in the ILA in 1976. I remain active in the women's movement nationally and internationally and I continue to engage in interdisciplinary scholarship that falls under the rubric of Black feminist studies; in this regard I continue to document and analyze the feminist theorizing and political activism of women of African descent; analyze gender politics within African American communities; and probe the history of Spelman College because of its unique mission of having educated Black women since 1881.

Most exciting at the present time is a project with Professors Stanlie James and Frances Foster which will be a new edition of *All the Women Are White, All the Blacks Are Men: But Some of Us Are Brave* which was the pioneering text on Black Women's Studies published by the Feminist Press in 1982. Our updated anthology, *Still Brave* (2008), will make visible the evolution of Black Women's Studies since the publication of *Brave* over 25 years ago. Before Professor Nellie McKay died, she stipulated that Frances and I should complete the updated version of this groundbreaking text with Stanlie since she knew she would be unable to do.

At a celebration of Nellie's life at the University of Wisconsin/ Madison in January of 2006, I wanted to remember those Black feminists who had impacted the field of Women's Studies, even though they often go unacknowledged: "In those early years ([1970s and 1980s]), we were a small band of renegade black women scholars determined to chart a new course within the academy and help to transform the two most transgressive spaces in higher education—Black Studies and Women's Studies. I'm speaking of Angela Davis, Nellie McKay, Barbara Smith, Barbara Christian, Mary Helen Washington, Hortense Spillers, Gloria Hull, Michele Wallace, bell hooks, Sonia Sanchez, Toni Cade Bambara, Alice Walker (who taught the first black women writers course at Wellesley College), Patricia Bell Scott, Frances Smith Foster, and many others."

In a recent note from Michele Wallace, who wrote the controversial *Black Macho and the Myth of the Superwoman*, she reflected on the difficulties that confronted those of us who early on claimed a Black feminist identity: "I can thank feminism for the fact that I am unmarried and childless but then a lot of women would curse it for the same reasons. I think we lack inspiring black feminist role models still—people who make black feminists (or other feminists of color) seem happy, fulfilled, content, and surrounded by love and understanding. The idea of spending your life growing wiser about human relations and gender is not the goal of most people, and certainly not must women, much less black women. On the other hand, to be a feminist intellectual of any kind, it seems to me the question of wisdom is central...But as a woman, wisdom isolates you from your peers and not in a good way.... We were and are and remain pioneers in a world in which it goes without saying that being male is the best of all possible existences. On top of which we have the disturbing issue of racial, cultural and ethnic difference."

I remain convinced that teaching antiracist, antihomophobic, anticapitalist, cross-cultural, transnational women's studies courses to undergraduate black women and men students is still the most important and pleasurable work that I do. Perhaps it is even the most radical, activist work in which I am still engaged.

In the aftermath of the twenty-fifth anniversary of the Women's Center at Spelman, which we define as radical feminist space, we celebrated the life and work of Audre Lorde. Through struggle, strategic resistance, perseverance, and clarity of vision about our work, we reflected upon not just the kind of program we had crafted, but about what often goes unnoticed. I'm referring to how we've been a haven for embattled junior faculty and staff; lesbian and bisexual students, staff, and faculty; scholar-activists from around the globe; gender-progressive men; artists, especially filmmakers attempting to complete their work, such as Aishah Simmons who finally finished her stunning documentary NO! which is about rape within Black communities; and literally jobless women who need a temporary place to call home.

Some have come with horrific narratives, including my own former secretary who was raped while her young son was forced to watch, or our middle-aged work-study student who had been homeless. Most have come with extraordinary gifts and hopes for a brighter future, such as Patricia McFadden, the African feminist scholar/ activist who participated in the antiapartheid struggle in southern Africa. She nurtures and feeds the minds and souls of our students in unbelievable ways; teaches an extraordinary course on African feminisms in a globalizing world; and speaks candidly to them about the joys of being a radical feminist who lives in the world on her own terms.

In a world that appears to be falling apart, where most of what we feminists care about is under attack, it is important to remember the joys of struggle and the power we still have as women's studies professors to impact the lives of young people in unimaginable ways. As the Women's Center prepared for our anniversary celebration, two of the many testimonies we received from former students attest to what is possible in these extraordinary women's studies spaces. Taneya Gethers, class of 2002, reflected on her time with us: "As a student, the Women's Center was a place where I was politicized and nurtured—and if I needed to vent and gather my sister allies, it was that place too. Love was never lacking. Sisterhood was always plenty. The Women's Center is where I, and so many others, became 'Afrafemme,' and we say thank you. We will always return, love deeply, and uphold the spirit of the program." Sarah Thompson, class of 2006, and the most radical student government president Spelman ever had, recalled: "I came to the Women's Center as a prospective student, and the Feminist Majority Leadership Alliance newsletter, 'The Gendered Lens,' caught my eye. As a high school senior, this was one of my first exposures to bold information about feminism, and my very first exposure to the words, intersectional analysis.

These two words, along with the words radical, African and feminist, flew into my vocabulary and my life when I entered as a first year student. The faculty there energized me to organize, to join the dance of activist movements, and to create safe spaces in which to describe—with new language—who I was and what I was about."[7]

The greatest testament to the power of women's studies for us at Spelman was the April 2004 protest by students (mostly women's studies majors and members of the campus-based Feminist Majority Leadership Alliance) against the negative portrayals of Black women in the rap music industry. Rap star Nelly was scheduled to come to campus in connection with a bone-marrow drive, but after students saw his music video, "Tip Drill," they demanded that he engage in a dialogue with students on the Spelman campus about his pornographic portrayals of young Black women. When he refused, they decided to hold a protest on campus anyway to bring attention to the more generic problem of misogyny in rap music videos.

In her June 4, 2004, article in the *Chronicle of Higher Education,* "It's Gettin' Hot In Here," Elizabeth F. Farrell had this to say: "Why did students at Spelman have such a strong reaction to the video? One explanation is that the campus has a unique environment: Out of 105 historically black colleges in the United States, Spelman is the only one with a women's studies department."

We certainly believe that the work of our women's studies program (we have courses such as "Women and Social Resistance Movements") and the feminist organizations on campus that our program has spawned have been major catalysts for the kind of feminist activism that the Nelly protest engendered. What we would never have been able to anticipate, however, is the national and international attention the student protest sparked, which included coverage by CNN, Fox News, Sunday Morning News, *USA Today, Washington Post*, a documentary (Byron Hurt's "Beyond Beats and Rhymes"), and several financial contributions from Spelman alumnae and other Black women.

Narratives of the development of Women's Studies almost never capture what faculty and students have been doing on Black college campuses. As Women's Studies continues to evolve and flourish, it is important to pause and reflect upon buried or forgotten histories. Without these alternative stories, and there are many more, it is nearly impossible to fully grasp how transformative Women's Studies has been in the lives of countless young women—including my own! It is also imperative that we imagine new futures for Women's Studies and ponder whether we are practicing what we preached nearly 40 years ago with respect to radically transforming the academy, improving the lives of women and girls, and bridging the disconnect between theory and practice. This is certainly what women's studies at Spelman College has been attempting to do since 1983.

Notes

1. Florence Howe's *The Politics of Women's Studies: Testimony from 30 Founding Mothers* (2000) is also instructive. Additional essays about my involvement in women's studies include: Guy-Sheftall, 1986; 1993; 1997; and 1994, which discusses my own feminist pedagogy; 2002, which discusses our Ford Foundation funded curriculum development project at Spelman on women and African Diaspora Studies.
2. Guy-Sheftall, 1979, 239.
3. The Combahee River Collective document is included in Guy-Sheftall, 1995b; see Ogundipe-Leslie, 1994.
4. This essay appears in *Feminist Politics, Activism and Vision: Local and Global Challenges*, eds., Luciana Ricciutelli, Angela Miles, and Margaret H. McFadden (London and New York: Zed Books, 2004, 100–121); this essay also includes an excellent bibliography.
5. Gillman, 2007; this essay also includes an excellent bibliography about the problematic of difference within mainstream feminism.
6. *SIGNS*, 32(4), 2007.
7. E-mail correspondence, September 2006.

References

Bell, R.P., B.P. Smith, and B. Guy-Sheftall. (1979). *Sturdy black bridges: Visions of black women in literature*. New York: Doubleday.

Braithwaite, A. and S. Heald. (2005). *Troubling Women's Studies: Pasts, presents, and possibilities*. Toronto, Canada: Sumach.

Farrell, E.F. (June 4, 2004) "It's gettin' hot in here." *Chronicle of Higher Education*, 50(39): A27.

Gillman, L. (2007). "Beyond the shadow: Re-scripting race in Women's Studies." *Meridians: Feminism, Race, Transnationalism*, 7(2): 117–141.

Guy-Sheftall, B. (1979). "Commitment: Toni Cade Bambara speaks." In R.P. Bell, B.J. Parker, and B. Guy-Sheftall (Eds.), *Sturdy black bridges: Visions of black women in literature*. New York: Doubleday, 230–249.

———. (1986). "Women's Studies at Spelman College: Reminiscences from the director," *Women's Studies International Forum*, 9(2), 1986: 151–155.

———. (1993). "A black feminist perspective on transforming the academy: The case of Spelman College." In Stanlie M. James and Abena P.A. Busia (Eds.), *Theorizing black feminisms: The visionary pragmatism of black women*. London and New York: Routledge, 1993, 77–89.

———. (1994). "Transforming the academy: A black feminist perspective." In David D. Downing (Ed.), *Changing classroom practices: Resources for literary and cultural studies*. Urbana, IL: NCTE, 263–274.

————. (1995a). *Women's Studies: A retrospective.* New York: Ford Foundation.

————. (1995b). *Words of fire: An anthology of African American feminist thought.* New York: New Press.

————. (1997). "Whither Black Women's Studies: Interview with Evelynn Hammonds." *Differences: A Journal of Feminist Cultural Studies,* 9(3): 31–45.

————. (2002). "Shifting contexts: Lessons from integrating Black, Gender, and African Diaspora Studies." In Mary M. Lay, Janice Monk, and Deborah S. Rosenfelt (Eds.), *Encompassing gender: Integrating International Studies and Women's Studies.* Old Westbury, NY: Feminist Press, 67–80.

Howe, F. (2000). *The politics of Women's Studies: Testimony from 30 founding mothers.* New York: Feminist Press, 2000.

Kennedy, E.L. and A. Beins (Eds.) (2005). *Women's Studies for the future: Foundations, interrogations, politics.* New Brunswick, NJ: Rutgers University Press.

Ogundipe-Leslie, M. (1994). *Re-creating ourselves: African women and critical transformations.* Trenton, NJ: Africa World Press.

Wiegman, R. (2002). Women's Studies on its own: A next wave reader in institutional change. Durham, NC: Duke University Press.

6

On Being a Pre-Feminist Feminist OR How I Came to Women's Studies and What I Did There

Evelyn Torton Beck

The Embattled Years

One could say I was born 'ahead" of my time, or into the "wrong" time, but whichever it was, I didn't fit. Having seen how traditional expectations of women had kept my mother from developing her keen intellect and thwarted her creative impulses, I knew early on that I did not want to live my mother's life. Nor, when I came of age in the 1950s, did I want any part of the rigid gender-bound dating arrangements of the time. I therefore made sure that the man I married understood that I had no intention of becoming "just a housewife." He agreed. We married. I earned a Master's degree and accidentally became pregnant. For him, that was the end of our agreement, but I neither could nor would let it go. After the birth of our second child, I forged ahead, enrolling in a Ph.D. program at the large state university of Wisconsin where my husband was a young professor. When I hired a babysitter to stay with the children while I attended classes, the wives in faculty housing talked about me behind my back and to my husband's face, but I was beyond caring.

Without my work, I knew I might end up mad—like Edna in Chopin's *Awakening* or the narrator of Gilman's *Yellow Wallpaper* (texts that had yet to be rediscovered). And because I knew these stories from inside myself long before I actually read them, I mustered a fierce driving force toward survival that gave me the courage to insist (over my husband's objection), on continuing toward my doctorate and a profession of my own.

In all the graduate seminars I ever took, we studied the novels of only one woman writer, Virginia Woolf, taught by the one woman in the department (Cyrena Pondrom), and even from her we learned nothing of Woolf's radical challenges to the patriarchy—*A Room of One's Own* and *Three Guineas*. Because this lone woman professor believed in my seriousness of purpose and respected my intellect, I trusted her guidance. It was she who finally became my dissertation advisor, rescuing me from the patriarchs in the department who believed I was working on a doctorate "as a hobby." This female mentor not only helped me find my scholarly voice, but also encouraged me to fight the university, when, after I had completed my Ph.D. and had a book to my name (*Kafka and the Yiddish Theater: Its Impact on his Work*, 1972), I was rejected by the Comparative Literature department for a teaching position that was being offered to a young male graduate student who had not even started his doctorate on the very subject of my book. Surprisingly, given the spirit of the times, an all-male committee found in my favor and I was offered a tenure track position. Although I won "my case," it is clear to me that without my mentor's support I might never have completed my dissertation or succeeded in breaking into the university world (which is what it felt like when I finally did get the job offer of assistant professor).

I never shared this "victory" with my Women's Studies students until I was getting ready to retire in 2002. In several decades of teaching Women's Studies, I had never questioned my reticence to mention this subject even when it was appropriate to the material we were studying. On reflection, it seems to me that because I had to fight my way in, my victory was tinged with some shame, when I should have been proud of my willingness to challenge the patriarchs in the department. It was my students, who, when they finally heard the story, helped me feel good about this act when they strongly applauded my actions and took comfort from my audacity in having taken on the institution, and won.

First Stirrings of Feminist Movement

But before I was on the job market, when I was entirely focused on finishing my dissertation while living abroad because of my husband's work, I was not aware of the stirrings of feminist movement until I returned to the United States and was greeted by the cover of *Ms.* magazine that featured Woman as Goddess with a dozen arms, each holding a different tool. The opening essay, "Why I Want a Wife," by Judy Syfers (later Brady) exposed the inequitable expectations of men and women within marriage and resonated with me so strongly I felt as if my world had just turned over. And indeed, it had.

The Women's Liberation Movement first took hold in my university as "Alice in Academe," an informal, off-campus program that offered noncredit courses taught by the volunteer labor of graduate students and a few junior faculty. The official Women's Studies Program was finally put into place only in the wake of prolonged protests when students (and a few faculty) occupied the university president's office.

Even though I was still years away from the job security that tenure brings, as part of a movement I no longer felt isolated, and as a result was willing to risk working with others for the transformation of knowledge and the development of the new interdisciplinary field of Women's Studies. In those still-embattled years, when so many feminist scholars were being sacrificed, I was able to maintain an aura of "respectability" as a scholar (even after I came out as a lesbian in my life and work) because my (pre-feminist) work continued to be seen as an important contribution to Kafka scholarship. For a time, I focused all of my teaching and research on women writers and artists, but soon, and with a feminist eye, I reapproached works of male writers like Franz Kafka (1982a, 1986, 1995), Heinrich Boll (1976), Isaac Bashevis Singer (1975, 1980), and others, in my research and teaching, thus bringing Women's Studies into the disciplines of German, Comparative Literature, and Yiddish.

In the years of burgeoning feminist activism when I also participated in marches and rallies both on and off campus, I believed Women's Studies was the academic arm of the women's liberation movement, and that I was part of an army of scholars engaged in the "culture wars." In the early years of its development, Women's Studies was denigrated and dismissed as trivial because of its foundational belief that "the personal was not only political," it was "academic" as well. Against heavy opposition, we fought to institutionalize Women's Studies to assure its survival in the university. Although decades later, it has come to present less of a radical challenge than in the early years of its formation, Women's Studies continues to challenge impugn masculist assumptions and methods, and its existence remains essential to the larger project of transforming knowledge and creating social change. Without Women's Studies and its allies in programs such as African American, Chicana/o, Asian American, Lesbigaytrans, the curriculum might well revert to its earlier versions and we could lose the gains we have attained. While I was director of the Women's Studies Program at the University of Maryland for nine years, I worked together closely with members of African-American Studies (the other programs had not yet come into being) in common cause. Later, many scholars from these programs also became affiliate faculty of what eventually became the Women's Studies Department that now offers an undergraduate major and minor, a graduate certificate, as well as a Ph.D.

What's In a Name?

I feel fortunate never to have been in a position where I had to fight to keep the name Women's Studies, which, more accurately, should have been called Feminist Studies in the first place. In those early years of formation, only at Stanford University did the program get away with using the adjective "feminist" (and it is not at all clear why they succeeded in that conservative institution). Years later, Stanford refused to tenure to its most prominent professor of women's history, Estelle Friedman, but eventually capitulated to national pressure that was brought to bear. There is no doubt that if most of us had tried to institutionalize something called Feminist Studies (at a time when "feminist" was equated with bra-burning, lesbians, and other negative images), we would never have succeeded.

But, after Women's Studies proved to be successful and drew many women and only a very few men, some programs strategically and voluntarily changed their names from "Women's Studies" to "Gender Studies," in the hopes that more men would sign up for classes and they would be more successful; at other institutions programs were pressured to accept such a compromise. While an analysis of gender is clearly central to the conceptualization of Women's Studies, I believe this name change is a mistake. In Gender Studies, women are subsumed under a rubric that may make men more comfortable while women are once again made to disappear. For similar reasons, I believe that Lesbian Studies should not be subsumed under Queer Studies and should remain part of Women's Studies. However, if, because of university politics, relocation becomes inevitable, at the very least Lesbian Studies should keep its name within the larger unit (whether it's called Queer or LGBT Studies) in order to maintain lesbian visibility.

Unity in Diversity?

I had quickly gravitated to Women's Studies when it was being founded, because it seemed clear to me that the liberation of women could only come about if the position and function of women—historically, culturally, and psychologically—were deconstructed so that women would become active agents in their own behalf and men would become allies because they would understand that there "are no free women until there are free men," as an early feminist poster proclaimed. In those early heady days, when we were trying to break the hegemony of the male subject ("he" subsumed "she"), many of us naively believed that we could meaningfully speak about "women" as a large general category that needed no differentiation. At the time this was both comforting and exhilarating, because if our experiences were the same, we could organize ourselves into a powerful force against patriarchy. But it very quickly became clear that this was a vision that had little basis in reality, as differences of socioeconomic class, race, ethnicity, religion, sexual orientation, dis/ability, geography, and nation, among others, divided women and had to be addressed in theory and practice, as well as in the curriculum. I came to understand that one could start by analyzing any form of oppression and if one followed it to its roots, one would see how all the oppressions were interlocking; for social change to come about, one could not simply focus on a single issue.

For me, coming to this realization meant I had to transform the new courses I had developed, such as *Women in the Arts* that at first had included only women writers and artists with whose work I was already somewhat familiar, not from formal education, but because of my own passionate interest in women writers and artists that predated feminist movement by many decades. Having been fortunate to grow up in New York City where museums were easily accessible, I had gravitated to the work of Kaethe Kollwitz (one of the very few women artists on exhibit at the Museum of Modern Art) where I spent many Sunday afternoons. I had also been a passionate reader ever since I learned to read English at age seven (having escaped Vienna and the terrors of the Holocaust reading was a real refuge), and already as a high school student had, on my own, found, and eagerly devoured the works of Kristin Lavrensdatter, Pearl Buck, Ayn Rand, Willa Cather, and other women writers in and among Dostoyevsky, Tolstoy, Camus, Stendhal, Knut Hamsun, and Alan Paton.

But almost as soon as I had put my first syllabus together, I was shocked when it became evident that those artists whose work I knew well enough to teach, all turned out to be white, middle-class, European or U.S. women. At the time, those of us developing Women's Studies courses were often only a few steps ahead of the students, learning as we were teaching, which was actually intellectually very stimulating and pedagogically fruitful. Because we too were reading and looking with "fresh eyes," we didn't presume to "know" the answers to the many questions that these texts evoked, and discussion was often at a high pitch of excitement as teachers and students together "discovered" women's voices even as we were developing our own. Knowing no better, and because lesbian lives were as yet unspeakable, we assumed that the artists we were studying were all heterosexual even if, as for Gertrude Stein, Willa Cather, Emily Dickinson, Muriel Ruckeyser, H.D., among others, later scholarship challenged this notion.

With so much new material rapidly becoming available, I soon developed courses in *Minority Women in the Arts* and *Lesbian Cultures* that allowed in-depth focus on artists and writers whose work stood even further outside the mainstream. I also created courses in *Women and the Holocaust* and *Jewish Women's Studies in International Perspective*, and over time recognized that no minority was singular, that each group was itself complexly diverse, and that overlapping and intersecting differences across groups were more common than rare. This recognition made it more difficult to create a curriculum for any given topic, as there was no way that a single course could ever be fully representative. This recognition sometimes made producing a syllabus painful instead of pleasurable; because I felt I was always leaving someone out, especially as new material on women on an international scope was quickly becoming available.

Balancing Sameness and Difference

Trying to balance women's sameness and difference proved to be a continuing challenge and one of the most difficult to address. The experience of trying to be inclusive led to the single most powerful insight I ever had as a teacher, one that transformed my teaching strategies, curriculum development, and choice of texts. I came to the realization that no single course I taught, no matter how "narrow" its focus, could ever be completely inclusive of all the differences, overlaps, and intersections that exist in the real world. For example, a seminar in *Lesbian Cultures* could always only include a study of selected lesbian cultures from around the world, but could not be all inclusive even if I focused only on Lesbian Cultures within the United States. When the topic was broader, such as *Healing Women: Feminist Perspectives on Mental Health*, my curricular choices became even more agonizing since I was determined to include the work of shamans and *curanderas* (traditional Latina folk healers) as well as feminist relational theories being developed by the Stone Center in the United States.

After feeling deeply discouraged, I realized that the sheer complexity of the multiplicity offered a way out. If one could not ever literally be all-inclusive in content, perhaps the answer was to offer a perspective that would bring to consciousness those who were not literally represented in the material. I would begin each semester by developing this perspective, hoping to provide a strong foundation that students would take with them as a lens into any course or body of material they were studying, whether or not it was Women's Studies.

The relief I felt in developing this strategy was enormous. No longer did I have to agonize for weeks about what I was "leaving out." It became clear to me that my task was to help students understand that whatever I (or anyone else) chose to include in any syllabus was, and always would be, partial and I urged students to become aware of what was *not* there, much as we had, in the early days of feminist movement, encouraged Women's Studies students to ask in mainstream classes, "Where are the women?" Similarly, they could ask in Women's Studies. "Where are the lesbians? Lesbians of color? Women of color? Jewish women? Disabled women?" And so on—the possibilities and intersections seemingly endless.

Nonetheless, I did not stop trying to broaden the dimensions of my syllabi, and I encouraged students to make use of term papers and research projects to expand the horizons of the course and share their work with the class. I continued to feel that what was absent should be named or become a felt presence in students' minds.

Digging Deep: Owning our own Assumptive Worlds

This strategy represented a kind of coming to consciousness that paralleled the consciousness-raising we did as women in the very early days of feminist organizing. While standpoint theory had become a significant tool in revealing the masculist biases of traditional scholarship and teaching, we had to go further and deepen the awareness of our own standpoints. For this reason, each semester, I started classes by asking students to bring to consciousness their own assumptive worlds, thus making visible the underlying assumptions they were bringing to the work. A few common examples that by 2007 may seem obvious (though still not to all students coming into Women's Studies for the first time): many students assumed that "woman" meant "white, heterosexual, middle-class, Christian female"; "mother" denoted a married (or at most divorced) heterosexual, biological female; "family" was constituted of biological parents and siblings. Until students became aware of their assumptions, they could not change their thinking. However, once the assumptions gave way, this awareness was eminently transferable to other concepts, and because it was not easily reversed, it usually was lasting.

However, this unearthing of assumptions was not limited to my work in the classroom. In order to help Women's Studies gain the professional credibility it needed in order to survive as a scholarly endeavor within the university, I became a founding member of the National Women's Studies Association (NWSA) and attended the founding convention in San Francisco. I was part of the team responsible for drafting the mission statement (of which we were very proud) and which included strong wording about the determination of NWSA to oppose sexism, racism, homophobia, class-bias, ageism, and looksism wherever it was found. Little did I know that several years later, after my own consciousness about anti-Semitism had been raised, the fact that we had not included anti-Semitism as part of our mission became a volatile issue, resulting in "difficult dialogues" and much unpleasant consciousness-raising within the organization.

Coming to consciousness has consequences we cannot foresee. I came out as a Jew in the lesbian feminist movement, after, to my shock and dismay, I discovered unacknowledged anti-Semitism within the movement. I was not willing to take risks on behalf of those lesbians who were denigrating my identity. As a child survivor of the Holocaust who had been helpless under the Nazis, I needed to take action. I joined with others to create both lesbian and Jewish caucuses within the NWSA. In an effort to work against the anti-Semitism I had discovered among lesbians and the homophobia I had found among Jews, I published *Nice Jewish Girls: A Lesbian Anthology*, a work that brought together these two identities (in all their complexity) and whose purpose was to educate both communities from inside (1982b/1984/rev.exp.ed.1989; update, 2007). A critical point of theory that grew out of this work was my recognition that "being a member of one minority did not necessarily keep anyone from being oppressive to others" and that "invisibility was itself a form of oppression."

From this theoretical basis, both caucuses worked to make NWSA's programming more inclusive. Over time, plenaries and sections on lesbians were added and became standard, but it was much more difficult to institutionalize programming, especially large plenaries on Jewish women. The Jewish caucus had constantly to remind NWSA leadership that we were invisible yet again. The most difficult, and for the Jewish caucus (most surprising), battle came when a proposal we had made to the governing assembly (including more than a hundred representatives), asking that anti-Semitism be added to the list of oppressions NWSA opposed, was met with harsh resistance and statements verging on blatant anti-Semitism that clearly elided Israeli foreign policy with all Jews. The caucus won that fight, but the stereotypes that emerged (some from Jewish women themselves) left us disappointed, with indelible scars and some mistrust of NWSA. This and other traumatic experiences in the women's studies world led me to write "The Politics of Jewish Invisibility" for the first issue of the *NWSA Journal* (1988), followed by "Jews and the Multicultural Curriculum" (1996), as well as a critique of the much used anti-Semitic and sexist epithet, "Jewish American Princess" (1991).

Over time, I firmly came to believe that oppressed groups must organize themselves before we could find the "unity in diversity" that was the goal of feminist organizing (1983b). But to this day, although only very partial unity has been achieved, some aspect of that initial vision—the "dream of a common language" among women across lines of difference—remains true: women do share some experiences based on gender, even while differences shape the specificity of those common experiences.

Self-Disclosure and Transparency as Pedagogical Questions

I had come to developing the discipline of Women's Studies from Comparative Literature with a great sense of relief. Although I had been trained in the years of formalism, when it was touted, following the poet Auden, that "A poem should not *mean*, but *be*," these approaches were never congruent with either my purpose or my way of teaching; my philosophy of teaching rooted the material in individual as well as social and psychological contexts; my pedagogical approach was interactive and relational. As a teacher, my purpose was to help students appreciate both the beauty of literature and its ways of helping us figure out our lives. Above all, like feminist poet Marge Piercy, I too wanted "to be of use." Women's Studies, with its foundational beliefs, "the personal is political," and the "academic is personal and political" offered me the opportunity to bring together the multiple dimensions of my life and work. The desire to be fully present and useful to the students' lives led me early on to consider how much of myself to share with the students, and when. This question became far more fraught once I came out as a lesbian and was teaching lesbian material in the classroom. It had not felt difficult to incorporate lesbian material into my classes when it was "they" and not "myself" I was talking about. Once I "was one," I no longer felt comfortable distancing myself from this denigrated minority, but was unsure of what it would mean to reveal myself. I wanted to serve as a role model, but was also fearful that I might lose my credibility. In contrast, and in spite of my work on anti-Semitism, which I knew still existed even on campuses, I was not uneasy in coming out as a Jew. It was a difficult, but not a "spoiled identity" in the sense that sociologist Erwing Goffman (1986) developed the term.

In addition to "coming out" in most of my classes, I also had developed a strategy that led to me make transparent, on the opening day of class, my intellectual history, how I came to develop the course, how I selected the material, what my approach was, and a quick overview of my life, past and present, which seemed most relevant to this class. However, over time, I began to worry that by offering students so much of myself, I wasn't in some way short-circuiting the development of their own perspectives.

I grappled with these issues in two essays written 12 years apart in two different teaching contexts. "Self-Disclosure and the Commitment to Social Change" (1983a) was the result of a small study in which I gave out questionnaires to students in my classes to see how they responded to my self-disclosure. I was pleased that most of the students, both in large lectures and small discussions, thought self-disclosure had positive effects: "it humanized the classroom, encouraged openness in students, created a good atmosphere, created unity in the groups, validated diversity and made the class more meaningful" (Beck, 1983a). I was also impressed that students came up with their own caveats, for example, teachers should self-disclose only when it is relevant to the subject matter, only if it doesn't take up too much time, only if it feels comfortable to the teacher herself. As a result of my research (mainly in the literature for psychotherapists—the issue hadn't yet been addressed in feminist pedagogy), I modified the timing of my coming out and the depth with which I introduced myself, and concluded that more important even than the actual act of disclosing, was the readiness to disclose and the integrative work it took to get to that place:

> ....such a stance toward ourselves, our students, and the material we teach creates a powerful synthesis where the point is not self-disclosure for its own sake, or for the sake of political correctness, but because telling seems important at a given moment, when it is most congruent with, and most organic to the act of teaching. (Beck, 1983a)

By the time I wrote "Out as a Lesbian, Out as a Jew: And Nothing Untoward Happened?" (1994), I had been visible for almost a decade as a very public representative of a Women's Studies Program intent on becoming a department (at the University of Maryland). To gain permission for new hires, new degrees (the major, the graduate certificate) and the approval of a Women's Studies Ph.D., I had been "outspoken, persistent, persuasive and not afraid of authority figures." I had also succeeded in raising a good deal of money. In this essay I grappled with the personal cost to me, my sense of self, to achieve all of this, because I was always aware that being both a lesbian and a Jew could result in others seeing me as a living embodiment of stereotypes. While I never compromised on the inclusion of lesbian material in our classes, and worked to get colleagues to include material on Jewish women, I did sometimes try to downplay my own lesbian writings. In rereading this essay, and with the benefit of hindsight, I realize that whether or not I included the subtitle to my book, *Nice Jewish Girls: A Lesbian Anthology,* everyone, even on so large a campus, knew anyway. Here, my memory of the difficulty we had in getting the NWSA to include anti-Semitism in its platform may have played a role in my discomfort in putting Jewish and lesbian together.

The Past is Not Dead

Faulkner was right when he said, "The past it is not dead. In fact, it is not even past." The building of Women's Studies not only remains amazingly vivid in my mind, but the basic principles of feminist consciousness continue to affect what I choose to do with my time, now that I have retired from university teaching.

About a decade before I retired, I decided to start work on a Ph.D. in Clinical Psychology (another early passion) through the Fielding Graduate University, and I was surprised to discover that although psychology had been one of the first disciplines subject to feminist critique and transformation, there was a great separation between feminist psychology and mainstream psychology. Although Fielding had as enlightened a faculty as I could realistically imagine, with many women professors, I found myself in a position vis-à-vis many of my professors uncannily similar to the one I had been in as an untenured feminist among masculist scholars in comparative literature. The psychology of women was not included, Jews were not part of their multicultural curriculum, and heterosexist assumptions were rife, especially among the male faculty. Because I could not keep silent, I helped to bring these omissions into consciousness, and what I hope will be a lasting contribution to the field grew out of these efforts. In collaboration with Julie Greenberg and L. Lee Knefelkamp, I coauthored an essay on "Integrating Jewish Material into the Teaching of Psychology" (2003) that gave a complex picture of Jewish identity across multiple lines of within-group differences.

But there was one area that I had unconsciously shied away from, even while I was expanding my awareness of what groups were still omitted from the curriculum—I always seemed to "forget" aging. Now, several years later, when I am about to turn 75, I find that I have finally brought together this theme with the realities of my life. After I completed this second doctorate (which had focused on the healing power of art in the life and work of Kafka and Kahlo, 2004), I realized that I did not wish to practice as a psychotherapist working with the troubled, but wanted instead to focus on developing well-being and creativity in the second half of life. I found myself gravitating to dance as a sacred practice, studying ancient women's dances from Eastern Europe that researchers believe contain messages from the grandmothers. The patterns our feet make resemble the patterns of these women's embroidery that I used to include in my earliest women's studies classes. What I saw in these patterns then (images of women's bodies), my teachers are seeing now.

Relying on my years of teaching, I began to offer workshops in sacred circle dances old and new—meditative movement to music from cultures around the world. Combining my training in group therapy, with my love of poetry, dance, and art, I am also offering workshops in Creative Aging and other inner journeys focusing on joy, peace, the seasons, health and healing, and women's development from maiden to crone. Once again I am learning as I am teaching, and feel thoroughly enlivened. Once again, as in the early years of Women's Studies, I find myself in a world of women, since very few men seem drawn to this dance practice or these workshops. My journey with women's studies has indeed brought me full circle.

References

Beck, E.T. (1972). *Kafka and the Yiddish theater: Its impact on his work.* Madison, Wisconsin: University of Wisconsin Press.

———. (1975). "The many faces of Eve: Women, Yiddish and Isaac Bashevis singer." *Working Papers in Yiddish and East European Jewish Studies*, No. 16, October. Reprinted in: *Studies in American Jewish Literature*, 1 (1981), 112–123. In Italian: *Comunita: Revista di informazione culturale.*

———. (1976). "A Feminist Critique of Boll's *Ansichten eines Clowns.*" *University of Dayton Review*, 12(2) (Spring): 19–23. Translated into German as: "Ein Kommentar aus feministischer Sicht zu Boll's 'Ansichten eines Clowns.'" In Anna Maria dell'Agli (Ed.), *Zu Heinrich Boll.* Stuttgart, Germany: Ernst Klett Verlag, 1984, 59–64.

———. (1980). "I. B. Singer's misogyny." *Lilith: The Jewish Women's Magazine* (Spring). Also in *Nice Jewish Girls* (1982, 1984): 243–249; reprinted in *The Jewish Socialist* 1:1 (Spring 1985).

———. (1982a). "Kafka's traffic in women: Power, gender and sexuality." *Newsletter of the Kafka Society of America*, 5(1) (June): 3–14. Reprinted in: *The Literary Review*, 26(4) (Summer 1983): 565–576 and Lazar, M. (Ed.). (1987). *The dove and the mole: Kafka's journey into darkness and creativity* (Proceedings of the Colloquia in Comparative Literature and the Arts). Malibu, California: Undena Press, 95–107.

———. (1982b). *Nice Jewish girls: A lesbian anthology,* Watertown, MA: Persephone Press; Reprinted, The Crossing Press, 1984; Revised and expanded edition, Boston: Beacon Press, 1989.

———. (1983a) "Self-disclosure and the commitment to social change." *Women's Studies International Forum*, 6(2): 159–163. Reprinted in Charlotte Bunch and Sandra Pollack (Eds.), *Learning our way: Essays on feminist education.* Thousand Oaks, CA: The Crossing Press, 285–291.

———. (1983b). "Unity in diversity." In *Selected papers from the 1983 New York State Women's Studies Association conference: Women in the '80s: Strategies for solidarity*, Albany: SUNY Press, 10–33.

———. (1986). "Kafka's triple bind: Women, Jews and sexuality." In A. Udoff (Ed.), *Kafka's contextuality.* Baltimore, MD. A joint publication of Gordian Press and Baltimore Hebrew College, 343–388.

Beck, E.T. (1988). "The politics of Jewish invisibility." *NWSA Journal*, 1(1) (Fall): 93–102. Reprinted in J. Butler and J. Walter (Eds.), *Transforming the curriculum: Ethnic studies and Women's Studies*. Albany: SUNY Press, 1991, 87–200. German translation in N. Kaiser (Ed.), *SelbstBewusst Frauen in den USA*. Leipzig: Reclam Verlag, 1994, 306–322.

———. (1991). "Therapy's double dilemma: Misogyny and anti-Semitism." In Rachel Josefowitz Siegel and Ellen Cole (Eds.)., *Jewish women in therapy: Seen but not heard*. New York: Haworth Press, 19–30. Also in *Women & Therapy*, 10(4): 19–30. Also see "From Kike to 'JAP': How anti-Semitism, misogyny, and racism construct the 'Jewish American Princess.'" *Sojourner: the Women's Forum*, 14(1) (September): 18–20, 1988. Anthologized in: H. Ehrlich and F. Pincus. (Eds.). (1994). *Race and ethnic conflict: Contending views on prejudice, discrimination and ethnoviolence*. Boulder, CO: Westview Press, 85–90.

———. (1994). "Out as a lesbian, out as a Jew: And nothing untoward happened?" In L. Garber (Ed.), *Tilting the tower: Lesbians teach queer subjects*. New York: Routledge, 227–234.

———. (1995). "Gender, Judaism and power: A Jewish-feminist approach to Kafka." In R. Grey (Ed.), *Approaches to the teaching of Kafka*. New York: A Modern Language Association Publication, 35–44.

———. (1996). "Jewish and the multicultural curriculum." In M. Brettschneider (Ed.), *The narrow bridge: Jewish perspectives on multiculturalism*. New Brunswick, NJ: Rutgers University Press, 163–177.

———. (2003). "Integrating Jewish material into the teaching of psychology," with L. Lee Knefelkamp and J. Greenberg. In P. Bronstein and K. Quina (Eds.), *Teaching gender and multicultural awareness: Resources for the psychology classroom*. Washington, DC: American Psychological Association, 237–252.

———. (2004). Physical illness, psychological woundedness and the healing power of art in the life and work of Franz Kafka and Frida Kahlo. Unpublished dissertation, The Fielding Graduate University.

———. (2007). "*Nice Jewish girls: A lesbian anthology. Revisited: 1982 and 2006*." In M. Grzinic and R. Reitsamer (Eds.), *New feminism: Worlds of feminism, queer and networking conditions*. Vienna: Löcker Verlag, 374–383.

Chopin, K. (1899). *The awakening*. Reprinted in various editions since 1982.

Gilman, C. (1973). *The yellow wallpaper*. Old Westbury, New York: the Feminist Press.

Goffman, E. (1986). *Stigma: Notes on the management of spoiled identity*. New York: Touchstone Press.

Syfers, J. (later Judy Brody). (1971). "Why I want a wife," *Ms*. December.

Woolf, V. (1929/1957). *A room of one's own*. New York: Harcourt Brace Javanovich.

Woolf, V. (1938/1966). *Three Guineas*. New York: Harcourt Brace Javanovich.

7

Women's Studies: Cultivating Accountability as a Practice of Solidarity

Ann Russo

If we ask ourselves the most simple questions, such as where do we get our food from, we can see that we are tied up in networks of relationships with millions of other people. Our actions are constantly creating, recreating, challenging and transforming the networks of relationships that make up the fabric of our shared world. We influence the fabric of society by the choices we make about whose actions we choose to acknowledge and whose we choose to ignore; by where we take a stand and where we choose not to; by how we treat others and how we expect to be treated.

(Kaufman, 2001, 33)

It's been 25 years since I first read Cherríe Moraga and Gloria Anzaldúa's *This Bridge Called My Back* (1981), bell hooks' *Ain't I a Woman* (1981), Angela Davis' *Women, Race, and Class* (1981), and Gloria Joseph and Jill Lewis' *Common Differences* (1981). These feminist writers rocked my world and transformed my feminist ideas, scholarship, and activism. They made me deeply question the underlying assumptions, limits, and dangers of different types of feminisms and feminists. They made me realize that feminist movement is not simply about fighting the outside enemy, but it is also struggling within myself and within feminist groups, communities, and institutions. They challenged me to be more reflective and introspective about my own identity, location, actions in relationship to and complicity in systems of oppression and privilege.

And it was in the context of conversations and debates within women's studies and feminist classrooms, programs, and activism that the ideas and truths became and continue to be real, relevant, and essential. It was in early 1982, in a Feminist Theories class, where I read these particular books for the first time. Our class erupted in conflict over the challenges posed by these scholars, writers, and activists. Their truths were made real to me as I listened to the defensiveness, anger, and guilt leveled against the writers and those of us who found them compelling. It's not that I didn't feel challenged, called out, and defensive, but I could see what they were saying as I reflected on my own experiences in feminist organizing and classes, and I identified with their anger at the injustices in and outside of feminism. My politics have never been the same as I experienced the refusal to consider our own limitations, failures, and complicity in other women's oppression. This experience radically shifted my thinking, my alliances, and my location in Women's Studies and in the world.

At the heart of this critical debate within women's studies is the question of how to theorize, politicize, and organize given that women are not a "natural" or "essential" group, that women do not share the same experience and oppression, and that women are complicit in systems of power and oppression. The difficulty has been that women's studies, like other identity-based academic programs, often has been conceptualized and developed from the assumption—at least in part—that women make up a coherent group with shared experiences and perspectives, a group that has been excluded, marginalized, and invisible in the academy and in the building of knowledge. The goals, in part, have been to analyze, deconstruct, and change the relations and structures of power. Power is often understood in terms of a binary—power/powerlessness, oppressor/oppressed—such that women's involvement in power structures and relations is often theorized from the position of powerlessness. The challenge has been to negotiate this simultaneity—a women's studies that centers women and yet does so without making women a homogeneous group, to center gender inequality simultaneously with other systems of inequality and privilege, to center issues of power and inequalities between women, and to insist on accountability as integral to the practice of and possibilities for coalition, alliance, and solidarity.

Over these many years, I have often returned to Cherríe Moraga's essay "La Guera," in *This Bridge Called My Back* (1981), where she reflects on the divisions between and among women and its impact on feminist solidarity. She attributes the ongoing divisions to the refusal of feminists to recognize our own implications in systems of oppression. She writes, "Within the women's movement, the connections among women of different backgrounds and sexual orientations have been fragile, at best. I think this phenomenon is indicative of our failure to seriously address ourselves to some very frightening questions: How have I internalized my own oppression? How have I oppressed?" (30). And therein lies the source of tension and resistance. For those of us in Women's Studies, my experience is that it is difficult sometimes to seriously consider how we ourselves may be responsible for and/or complicit with oppression, discrimination, and violence through our words, actions, theories, classrooms, and practices. These issues of difference, inequality, and accountability, for me, continue to be at the heart of women's studies and feminist scholarship, teaching, and organizing today.

The recognition that our work within Women's Studies may be complicit in, at the same time that it is resistant to, oppression dislodges a simple formulation of us/them—oppressed/oppressor, victim/perpetrator, powerless/powerful—and moves us toward a serious analysis of the enmeshment of systems of oppression *and* systemic privilege in scholarship and teaching. It challenges us to consider our own participation in these systems of power. This shifts the discussion away from erasing our differences, or comparing oppressions, or creating a hierarchy of needs, and toward an interrogation of our complicity in these systems whether we are oppressed by them or not. This process is what makes solidarity across difference possible and leads to expansive feminist frameworks and politics. Although this is often a difficult process, as evidenced in my classrooms and in women's studies conferences and meetings, it can also be transformative. Emi Koyama (2001), in her argument for "trans women to openly take part in the feminist revolution," writes, "Every time a group of women previously silenced begins to speak out, other feminists are challenged to rethink their idea of whom they represent and what they stand for. While this process sometimes leads to a painful realization of our own biases and internalized oppressions as feminists, it eventually benefits the movement by widening our perspectives and constituency" (1).

Interlocking Systems of Oppression and Privilege

> On the one hand if we understand the realities of groups subordinate to us as different or special, we plunge into hierarchy: we become saviours of less fortunate peoples. On the other hand, if we start from the premise that non-disabled people are implicated in what happens to women with disabilities, we might stand a better chance of detecting when we are simply re-establishing our superiority by noticing difference. (Razack, 1998, 20)

Within the field of women's studies, the perspectives of feminists of color, lesbians and bisexual women, transgendered folks, poor women, "third world" women, women of the global South, women with disabilities have shifted feminist theorizing to consider the multiplicity of identities and structures shaping this loosely constructed group called "women." Writers such as Alexander (2005), Allison (1994), Moraga and Anzaldúa (1981), hooks (1981; 1996), Hurtado (1996), Koyama (2001), Mohanty (1991; 2005), among others, have offered social analyses that consider the many interconnected systems of oppression and privilege (e.g., patriarchy, white supremacy, capitalism, compulsory heterosexuality, ableism, colonialism, imperialism) in terms of how they differentially shape people's lives, ideas, relations, and politics. These writers underscore the significance of accountability in terms of whether women's studies theories, research, and politics reproduce and/or resist dynamics, relationships, and systems of privilege and oppression.

These conceptual frameworks of interlocking systems of oppression and privilege are the basis of many women's studies classes, anthologies, conferences, and organizations. There are a variety of ways that these frameworks are articulated, and yet the basic idea, as Fellows and Razack (1997–1998), suggest, is one in which we understand that "systems of oppression (capitalism, imperialism, patriarchy) rely on one another in complex ways. This 'interlocking' effect means that the systems of oppression came into existence in and through one another so that class exploitation could not be accomplished without gender and racial hierarchies, imperialism could not function without class exploitation, sexism, heterosexism, and so on" (335). This means that all of us are shaped by and implicated in these systems—our identities, our experiences, our ideas, our actions.

Although these frameworks have become a kind of mainstay within Women's and Gender Studies, it seems to me that there is still a tendency to mostly focus—in theories, research, classes, and programming—on one or two axes of oppression, and in the rare instances when privilege is considered simultaneous with oppression, the focus tends to be on race privilege (usually "white privilege"). I struggle in my classes to cultivate a sustained and simultaneous focus on the forces that shape how privilege, access, and power shape identities and experiences of oppression and resistance, as well as *complicity* in others' oppression. I often draw upon the words of Elizabeth Hobgood (2000) who describes the dilemmas of privileged class women; she writes, "Insofar as we occupy multiple social locations, the lack of power and freedom we experience from our membership in subordinate groups may blind us to the power we have as members of dominant groups. Our moral situation is complex. We must acknowledge both our pain and our privilege, both how we are constrained and where we have power, if we are to attain responsible moral agency" (10). And this simultaneity of oppression and privilege is one facing Women's Studies itself as a field given how it has grown and developed in its own academic legitimacy and place in the academy. As faculty members and also as administrators, our levels of relative power and privilege have increased. Although many of us continue to struggle from positions of unequal power with respect to white male hierarchies, we have our own institutionalized hierarchies within Women's Studies programs and institutions for which we are also accountable.

Although there has been some change over the past 25 years of my involvement in Women's Studies, it remains true that Women's and Gender Studies continues to be centered in normative U.S., white, middle-class, heterosexuality. This is evident often in how issues of power and privilege get played out in organizations such as the National Women's Studies Association (NWSA) and also within local programs and organizations. The majority of the leadership in Women's Studies continues to be white and/or middle-class and/or U.S. and/or heterosexual academics, and we are the ones who often have the power in determining agendas, canons, curriculum, and knowledge-formation. The issue is not whether or not those of us are privileged by race, class, ability, citizenship, sexuality, and so forth should be in the field and participate in decisions; the more significant question, I think, is how do we understand our role and our accountability with respect to the inequalities and structures that produce our power and leadership in the production of knowledge and its consequences (see, for instance, Koyama, 2000).

From my experience and through my participation, it seems that faculty and students alike often are willing to embrace the idea of "difference" and even recognize "inequality" among and between women, but are more reluctant to examine our own integral relationship to how these differences and inequalities are produced. And yet this is important; as Hobgood (2000) suggests, "critical analysis of these systems [must help] us see how privilege and oppression do not simply coexist side by side. Rather, the suffering and unearned disadvantages of subordinate groups are the foundation for the privileges of dominant groups" (16). In a white supremacist capitalist heterosexist patriarchy,[1] women exist in hierarchical relation to one another, and our lives are connected through these hierarchies.

This kind of analysis, for instance, prompts me to reflect on how these hierarchies get produced within the field in terms of how the core curriculum gets established, new faculty lines are conceptualized, and/ or key programmatic goals are realized. It also makes me reflect on how demands for more resources in women's studies program impact the labor of other women and men in the university (e.g., the resulting increased labor of the clerical, operations, and other staff at the university whose voices, labor, and compensation are not always figured into the equation). And it encourages me to think about how discussions about the status of women in the university often do not include the women and men doing this labor. And it makes me think about how calls for women's solidarity across the university may be reflections of a power relationship, not a relationship of sisterhood based in shared oppression. In exploring the possibilities for creating coalitions among and between women, Cricket Keating (2005) emphasizes that "building feminist solidarity requires critical self-reflection that acknowledges that how one lives impacts the lives that others are able to live.... we are all connected because of our relational insertion into hierarchies of power and privilege, hierarchies that we also can resist and transform" (93).

Tracing Complicity, Taking Action

Much anti-oppression work within women's studies has been to recognize "difference" and to include "other" women in programs, scholarship, and politics. However, I think these efforts do not necessarily challenge the ways in which we are structurally related to one another. In order to shift away from both the "shared oppression" and the "difference" frameworks, I have begun to think more in terms of tracing people's involvement in systems of oppression and privilege. In the context of white antiracist practice, for instance, Leslie Roman (1993) offers that white people must locate ourselves "*in* the stories of structural racism," rather than outside of them. For instance, we need to create narratives where we "fully account for the daily ways we (whites) benefit from *conferred* racial privilege as well as from our complicity in the often invisible institutional and structural workings of racism" (84). Such tracing might allow for more action and movement in relation to these systems of power that we might otherwise see ourselves as outside of. By encouraging faculty and students in women's studies to recognize our structural relationship to one another and our involvement in maintaining these systemic hierarchies, I think we might, as Sherene Razack (1998) suggests, move from a *politics of inclusion to a politics of accountability.* She writes that if we recognize how we are implicated in the "subordination of other women" then our "strategies for change" will "have less to do with being inclusive than they have to do with being accountable" (159).

In my classes, I often draw on my experience a few years ago in an antiracism training for an LGBT group in Chicago called Queer White Allies Against Racism. The trainers had us watch a short film, *Space Traders,* based on a short-story by Derrick Bell (1992), from his book, *Faces at the Bottom of the Well.* Briefly, the story is as follows: an alien group lands in the United States and offers the following: "Give up all African-Americans and in return they will leave for America the contents of their ships—enough gold to retire the national debt, a magic chemical that will cleanse America's polluted skies and waters, and a limitless source of safe energy to replace our dwindling reserves. Following roughly two weeks to consider the offer, heated public debate leads to a referendum that when approved, consigns all Blacks to the traders" (Bell, 1999). As the story develops, almost all of the white leadership and the majority of white people (70%) vote for the tradeoff on the basis of its benefits for them; some even frame it as patriotic for Black people to do this service for the nation because it would restore all natural resources to the United States. Although there are efforts to organize people of color and liberal whites to resist the resolution, they are ineffective.

After showing the film, the trainers asked our group of queer white allies: who did you identify with? Most of us reported that we identified and empathized with one or more of the African American characters in the film—we related to their struggle, the dilemmas, the pain of being betrayed by friends, singled out, targeted. None of us spoke about the white people in the film; instead, most of us distanced ourselves from them and from their decisions. The trainers challenged us to shift and expand this lens. They suggested, instead, that to become allies for racial justice we, as white activists, needed to see ourselves in the stories of the white folks in the film and to consider how we may be implicated in this story. We needed to ask ourselves—what is our connection with the current state of race relations in this society? Who are we in this picture? How are our own decisions, in the groups and institutions to which we are affiliated, connected to those of the various white people in the film? In other words, as white antiracist activists, in addition to empathy, the trainers challenged us to recognize ourselves in the ideas and actions of other whites, to take responsibility for our power and its possibilities for resistance. They asked us to explore our implication in the racism and white supremacy of this society, rather than to place ourselves outside of it or as victims of it.

I encourage such discussions in my classes. It offers a concrete way to deeply engage in the realities of our own active and passive involvement in social justice issues. By forcing us to seriously consider our own participation and to take responsibility for the consequences, we are all able to make more intentional choices about our actions and our alliances. I think this is what is often difficult for many of us to admit—the ways in which we are very much a part of one another's lives and contribute to what happens to one another. We often seek for a way out of this responsibility by either seeing ourselves as "the same" or as completely "different." And yet, it is so clear in terms of working with students and other antiracist activists that this recognition often frees us up to act more and to let go of the desire to be "innocent" or seen as "the same." Insisting on a "shared oppression" model of solidarity doesn't free us up to act, to see clearly what kind of role we have in challenging these practices and systems.

Resisting Amnesia

One of the barriers to engagement in this work is the lack of historical identification among white middle-class people in the United States. In my classes, for instance, white students often struggle around identifying their "people." Mostly, they do not have a sense of how to place themselves in history, and they often resist seeing themselves connected to histories of oppression, violence, and domination. In my antiracist feminism course, I try to encourage all students, and especially white and middle-class students, to resist the historical amnesia about their own people's entry into and involvement in history, particularly as it relates to histories of oppression, privilege, and resistance. We read, for instance, Mab Segrest (1994) and Aurora Levins Morales (1998) who provide great models of how to understand and own these histories and legacies of inequality, colonization, slavery, white supremacy, capitalism as well as legacies of resistance and change.

I often use Aurora Levins Morales' (1998) book, *Medicine Stories*, where she offers a powerful methodology of owning complex, contradictory, and painful histories of oppression and resistance. She shares her own story of coming to terms with the complexities of her family. While she grew up with stories of her mother's family's struggles with poverty and racism in the Bronx, New York, through her travels to Puerto Rico, she found out that she also comes from "five generations of slave-holding ancestors among the petty landed gentry of northeast Puerto Rico" (76). She talks about the shame, upset, and discomfort she experienced when she learned about her family's history, and yet she writes, "If I could figure out how to face it and consciously carry it, how to transform shame and denial into wholeness, perhaps I could find a way out of the numbness of privilege, not only for myself, but for the people I worked with in classes and workshops who came asking to learn" (76).

Morales (1998) offers an important model of how we might cultivate accountability for own particular historical legacy; she, for instance, uses her family history to "break silence, to acknowledge publicly and repeatedly my family debt to their coerced labor, to expose and reject family mythology about our 'kind' treatment of slaves..., and to make a decision that although none of these people had chosen me as a descendent, I owed them the respect one gives to ancestors B/c spell out because their labor had made it possible for my forebears to grow up and thrive. I have also made it my responsibility to make African people visible in every discussion of Puerto Rican history in which I participate" (76). Morales (1998) offers students and faculty a way to see how taking responsibility can actually feel "empowering and radical." She writes, "Guilt and denial and the urgently defensive pull to avoid blame require immense amounts of energy and are profoundly immobilizing. Giving them up can be a great relief. Deciding that we are in fact accountable frees us to act.... [It] leads to greater integrity and less shame, less self-righteousness and more righteousness, humility and compassion and a sense of proportion" (76).

In the last couple of years, I've had the opportunity to teach feminist histories. Given that these histories are also mired in women's power, oppression, and resistance, I've found myself drawing upon Morales' approach here as well. Over the past 30 years, feminist historians have contributed a substantial body of research that analyzes the involvement of earlier feminist scholars, activists, organizations, and movements in racist, eugenicist, imperialist, classist, and homophobic projects (e.g., Ahmed, 1993; Chaudhuri and Strobel, 1992; Newman, 1999; Ware, 1991). Although the writing of feminist histories often began as reclamation projects to honor and celebrate "our" past, this is simply unacceptable given the enmeshment of feminists in these legacies of oppression and violence. The question is how to teach these histories, how to be accountable to their consequences, and how to talk about them in relation to contemporary feminist theories and politics. In my class, we struggle over how to understand these legacies. We talk about the importance of facing how various feminist individuals, groups, and movements negotiated, reproduced, and resisted oppression and privilege. For some this proves quite daunting; some do not want to face or want to justify the complicity of Margaret Sanger's arguments for birth control, for instance, with eugenics campaigns, or the racism and classism used by white middle- and upper-class women's suffragists to gain support for the (white) women's vote. Other students feel betrayed and angry that these women have been held up as heroines of feminism. It is important to negotiate the complexity of these women's stories—the ways in which they resisted oppression and the ways they reproduced it.

In an effort to encourage students to recognize multiple ways of negotiating these feminist theories and histories, I sometimes share my own history of reading these texts. When I first began researching the late-nineteenth and early-twentieth-century U.S. women's rights movement in the early 1980s, for instance, I didn't know how to deal with the racism, anti-Semitism, xenophobia, and classism in the speeches and writings of many of the most prominent women featured in feminist histories. Sometimes I would try to minimize these aspects of their work. I wasn't sure how to think about what I was reading and so I tried to see these aspects of their work as less significant than the parts that seemed more resistant and radical. Other times I would become infuriated with their ideas, and I would distance myself from these women, and decide they were not worth reading. In the latter case, I would sometimes reassure myself that these women were not "my" women, given my own family's background as Irish, Austrian, and Italian working-class and poor Catholic immigrants who were not embraced by some of these nineteenth- and early-twentieth-century advocates for women's rights. I talk with students about how both of these strategies were problematic (which many students can see once I describe them) and then I share with them other ways I've found of thinking about these legacies and my relationship to them. I talk about how I recognize that these feminists, in part, with all of their limitations and problems, are still very much connected to my own life possibilities including the ways I initially embraced main-stream feminist ideas. The gains made expanded the possibilities for white middle-class, and sometimes working-class, women in the United States, including those in my family, often on the backs of or in opposition to the rights and gains of women and men of color. I trace how contemporary feminist ideas have often uncritically repro-duced the racism, classism, homophobia, ableism, and imperialism that were integral to many of their ideas. I believe that critical and honest engagement of these legacies is one way of taking responsibil-ity and of recharting the paths of contemporary feminisms toward more justice and accountability (e.g., Bannerji, 1997; Davis, 1981; Newman, 1999). Following the guidance of Dorothy Allison (1994), we need to tell the "messy" stories and interrogate them for the ways that may reproduce and/or resist oppression so that we might speak more truthfully about where we've been and hopefully provide more clarity about the paths that we are now forging.

These issues are also relevant to teaching about the development of contemporary feminist theories and politics, including the his-tory of Women's Studies. Many feminist scholars and activists have uncovered and critically interrogated the ways in which the field and its practitioners have been complicit in fueling multiple oppressions through its institutionalization within the academy, its research and scholarship, and its pedagogy based in assumptions about women's "shared oppression." These critiques have helped to transform many aspects of women's studies, and continue to be areas of contention and change. For instance, one can trace these struggles within the NWSA. Since its inception, NWSA has faced ongoing internal strug-gles around racism, classism, homophobia, ableism, ageism, and other forms of oppression. I have been a part of these struggles, for a few years in the mid-1980s while I was a graduate student, and then since 1995 when I became a full-time faculty member.

My experiences in NWSA have been both profoundly frustrat-ing and sometimes very hopeful. On the one hand, NWSA has not lived up to its mission of anti-oppression and full inclusion; on the other hand, it has a long history of resistance by its members to make it more accountable to its mission. In the past several years, I have had the opportunity, through the Anti White Supremacy Taskforce (AWSTF) and the Women of Color Caucus (WoCC) and the Stop Dreaming, Keep Working annual gatherings, to work with others in NWSA to address institutionalized racism and white supremacy in the organization by building an antiracist and social justice community within NWSA. We have had, and continue to have, some very powerful moments of collaboration and alliance building (see Russo, 2005), and we've also faced real resistance and animosity. Organizational change is hard, slow, and often fraught with tension and conflict. It's hard to build community and connection, and there are many forces that make it difficult. And yet it is possible and I've seen it happen.

One of the issues, then, has been how to characterize NWSA and these ongoing dynamics of oppression and resistance. The history of conflicts around racism and white supremacy (among other oppressions), for instance, is often evaded, or present only in whispers and side comments. For example, the infamous "Akron conference" continues to haunt the organization. It was at this conference that a significant group of women of color and white allies walked out of the conference and out of the organization due to entrenched racism. Since my return to NWSA in 1995, this event is often briefly referenced as a "terrible moment" in NWSA's history. It is rarely explained in terms of racism in the organization. The issues at the conference and the deep impact on the organization are seen as caused by the walkout, not the racism that preceded it (see for instance Koyama, 2000). The organization continues to struggle to find a way to own the history and create accountability for the ongoing issues of mistreatment, inequality, and institutionalized oppression. At the 2001 conference, the WoCC and the AWSTF worked together to demand, and then collaboratively develop, a plenary on women of color and the issue of racism at NWSA; the plenary was entitled, "Women of All Colors Building an Inclusive Organization Together," with Beverly Guy-Sheftall and Lisa Albrecht (Aida Hurtado was scheduled but unable to present because of illness). We also cosponsored a session to create a conversation around the 1991 "Akron conference." These events served as another effort to create institutional ownership and accountability for the events of the Akron conference, and for the struggles that preceded and followed it. And the struggle continues.

NWSA—like many feminist groups—is complex, contradictory, and in process. I often wish it was different—more progressive, more inclusive, more accountable, more oriented toward social justice. At the same time, I am very much a part of its history of internal resistance and struggle. Following the lead of Aurora Levins Morales (1998), I think it's important to own NWSA's conflicted history and to honor the work of all the women of color, lesbians, bisexuals, queers, transgendered folks, women with disabilities, women of the global South, old women, allies, and/or others who have struggled, resisted, and challenged NWSA, as well as those who left the organization in protest. In the past few years, I've encouraged students in Women's and Gender Studies to attend the conference. Inevitably, many complain to me—they are disappointed and frustrated that it does not live up to its mission or the ideas embedded in much of what they read in their classes. I urge them not to give up on the organization—to realize that no organization is "pure" or "perfect" and such standards cannot be the measure with which to gauge a decision to be involved or not. The question, for me, continues to be how to continue to build an NWSA—and a politics—that will be more accountable for its history and present—not by ignoring the negative, or minimizing the oppressive structures and practices, or erasing the ugly and mean stories, and/or, in contrast, just giving up on it. Instead, I continue to hope that those of us who are interested in this struggle will continue to find ways to cultivate communities of resistance and transformation within the organization. Institutional struggle is difficult and often painful, and yet necessary and sometimes transformative.

Refusing Innocence

> All of us have had failures of integrity. I believe part of what makes it so hard to consider perpetrators as part of our constituency is that we cannot bear to examine the ways in which we resemble them (Morales, 1998)

One of the most powerful barriers to owning up to our own implications in systems of oppression is the desire for "innocence." Although the theories and analyses we draw upon in Women's Studies mostly recognize that oppression is systemic, structural, and embedded in all of us, we often respond as if it's mostly about individuals. Thus when we are challenged to be accountable for the perpetuation of someone else's oppression, we often resort to defending our intentions as innocent. Again, it's difficult to own up to how we are connected to the subordination of others. Fellows and Razack (1997–1998) discuss how such responses ultimately reproduce oppression; they write, "When we view ourselves as innocent, we cannot confront the hierarchies among us" (338). The fixity of the positions of oppressed and oppressor continue to stifle the possibilities for accountability across lines of difference.

I see this in my classes and in my work within the university. It's often easier to draw rigid lines between oppressor and oppressed and to place ourselves with the oppressed. There's a reluctance to examine our interrelatedness with the privileged and powerful in a system of oppression, and yet this is an important space to self reflect about complicity and to engage in resistance. In recent years, this is what I've been trying to develop in my activist scholarship as well as in my classes. I seek to create more conversations, for instance, in my women and violence class around community accountability. I encourage students to move beyond an us/them, victim/perpetrator approach to the issues. I talk about the reality that all of us are members of communities and, in the context of the many forms of violence in and across these communities, we are all accountable—survivors, perpetrators, bystanders, among others—for the perpetuation of violence. Thus, we are all responsible for engaging in efforts to end oppression.

I've begun to reference a story in Thich Nhat Hanh's (1987) book *Being Peace*. He tells the story of a group of sea pirates who rape a 12-year-old girl. I talk about my own response to the story—a response that he too expects. When I read the story, my heart goes out to this young girl who has been viciously raped and I hate the rapists. Hanh, however, like Bell, challenges me to see my relationship to the sea pirates. He suggests that in order to stop rape—we must consider the conditions that create the sea pirates, or any others who commit violence, rape, murder. No one is born a rapist. The question, then, is how did they become rapists? Or bigots? Or racists? This does not mean taking away responsibility from the person who has perpetrated the violence; it does mean, asking why this violence happens and asking how such violence is produced by the communities in which we live.

I think often of an experience I had many years ago. I was teaching a course on violence against women, and I had a young man in my class that was relentless in his critique of feminism and his anger at and blame of survivors of sexual assault. As an adjunct teacher with little experience, I didn't know what to do. I was afraid to kick him out, and I had trouble getting him to stop. At the last session of the class, he again berated me and the class. I finally asked him to stay after class so that I could tell him about the harm that he had caused by his hostile attitude and words. He stayed after the class and began pouring out his anger. I listened. Within five minutes, we were sitting down and he began to tell me about his own history as a survivor of sexual assault and how bad he felt about himself. He talked of suicide, of more punishment. We talked about ways to understand and address these feelings and I shared some community resources that might assist him as a male survivor of sexual abuse. I'm not excusing his behavior in the classroom, and yet, when I'm faced with similar young men and women in my classes, I try to take time to listen to what lies underneath their resistances. Through this listening, I always learn more. And sometimes, in the process, I am able to work with them to carve a new path for us that restores community. This new path recognizes that accountability is essential and, in part, comes through a recognition of all the conditions that brought us to where we are in that moment.

As a survivor of sexual assault, I could refuse these questions and see myself as innocent, not involved, not responsible for resistance and/or violence. But this refusal is no longer acceptable to me. In Adrienne Rich's (1986) poem, "Virginia 1906," she meditates on the power of a white woman who has been raped, challenging us to not use our violation as a mask of innocence in the face of the violation of others. She asks, "I am tired of innocence and its uselessness,/.../ Nothing has told me how to think of her power./.../Do we love purity? Where do we turn for power?" (41). Along these lines, within the context of Women's Studies, I think we need to ask ourselves: how does a racist, classist, imperialist, ableist, or transphobic feminism get developed? What is our role? What accounts for the reproduction of oppressive ideas, practices, structures, and actions? What are we doing to transform feminisms to make it more accountable, powerful, expansive?

In my classes, I have begun to use Cricket Keating's (2005) method of "coalitional consciousness-building" that she developed based on the work of Lugones (2003), Mohanty (1991), Reagon (1998), and others. It's a process similar to consciousness-raising, but one that seeks to draw out, rather than minimize, "the multiple relations of oppression and resistance at play" and that explores "the barriers to, and possibilities for, coalitional action..." (Keating, 2005, 94). Rather than building on monolithic ideas of unity based on shared experiences and oppression, the process assists in identifying common issues with attention to different relations to power and privilege within these same issues. She offers a three-step process for groups to explore their experiences in ways that assume differential relationships to power, different possibilities for resistance, and different locations of oppression and privilege in relation to the situation at hand. This process allows students to hold on to their hope of coalition and alliance across differences and inequalities at the same time that it makes all of us more accountable to one another given our different power and resources. Recognizing the places of one's involvement creates many opportunities for more explicit action. It offers all of us places to reflect and act in recognition of the areas where we are privileged, by race, sexual identity, class, and/or citizenship, and so on. It provides a method of discerning more opportunities where we can step up to the plate, act, and be a participant, not a bystander, in the struggle to create more just communities within and outside of feminism and women's studies.

Exploring, naming, and claiming one's privilege is an important part of the process of anti-oppression work (Kimmel and Ferber, 2003; McIntosh, 1988), and students in my classes wrestle with their own privileges. One of the stumbling blocks, however, has been that many seem to get caught up in the question of individual identity and feel unclear how to move forward with their knowledge. This is often due to the ways in which identity is often conceptualized as static and unchanging. For instance, while they are able to now say—"I am white, I am middle-class, and I am privileged in these ways," they feel that they are stuck in this identity that they now feel is problematic. I often hear students say they feeling either demoralized, or guilty, about their own situation of privilege, and don't see how to use this knowledge as a space for resistance and action. I often refer students to Barbara Smith's (1984) clarity and moral challenge where she writes

No one on earth had any say whatsoever about who or what they were born to be. You can't run the tape backward and start from scratch, so the question is, what are you going to do with what you've got? How are you going to deal responsibly with the unalterable facts of who and what you are, of having or not having privilege and power? I don't think anyone's case is inherently hopeless. It depends on what you decide to do once you're here, where you decide to place yourself in relation to the ongoing struggle for freedom. (77)

And it is in placing ourselves in the "struggle for freedom" that becomes an opportunity for changing who and what we are. I am very appreciative of the work of Aimee Carrillo Rowe and Sheena Malhotra (2006) who are theorizing the "unhinging" of whiteness and heterosexuality from static identities and locations to offer more possibilities for resistance. Aimee Carrillo Rowe (2005) illuminates how our identities can shift and transform through our choices of connection, community, and belonging. She offers a vision for coalition and alliance that speaks to our yearning for community and connection across lines of difference and power. She suggests, "who we love is political. The sites of our belonging constitute how we see the world, what we value, who we are (becoming)" (16). She reframes the "politics of location from the individual to a coalitional notion of the subject" (16), that is relational rather than stable and unchangeable. Building on Chela Sandoval's (2000) "differential consciousness," Carrillo Rowe posits the notion of "differential belonging" across borders of race, class, and sexuality. In creating relationships and ties across these borders, we also change who we are, what we know, and how we act. She writes, "Differential belonging calls us to reckon with the ways in which we are oppressed *and* privileged so that we may place ourselves where we can have an impact and where we can share experience.... When we place ourselves with people aware of their oppression, we begin to see how we are implicated, to wrangle with the connections between privilege and oppression, not as abstract concepts but as constituting 'our' lives" (35). She locates the impetus to action and change in terms of our relations with one another, rather than solely in terms of our own identity and self-knowledge (which are never really individualist or isolated in the ways we are encouraged to believe).

As scholar-activists engaged in movement, our identifications, relationships, communities, and loyalties are not stagnant and through our choices and our actions, they shift. Mine have shifted dramatically. If I lived out the plan laid down in my conservative Roman Catholic upwardly mobile middle-class white family in the late 1950s and 1960s, I most certainly would not be living the life I am living, making the choices I am making, and building the relationships and communities that I am building. The path I am forging is one that I continue to develop—in struggle and in opposition to how I grew up and with a deep yearning for connection and community in a "remade world" (Allison, 1994). And, of course, the choices I have made are connected to both pain and privilege, both—not one or the other—but both. I recognize the privilege and power of my location in the world and try to be accountable to it in the work with others for social justice.

Last spring (2007), when activist-theorist Suzanne Pharr visited DePaul University, she talked with a group of students and faculty who were trying to address hate speech on campus. She encouraged us to think about building a politics that mobilizes around our yearning, as much as focusing on what we are against. I was reminded of this in the fall (2007), in my Feminist Theories class, when students expressed frustration with the difficulties and barriers to build unity in groups and movements across differences. I wanted them to understand why their insistence on "sameness" and "unity" evaded their own implication in the oppression of other women. In this interchange, however, I realized that they had never seen or experienced a context in which people grappling together with difference and inequality contributes to the building of community. bell hooks (1996), in *Killing Rage, Ending Racism*, talks about the importance of making more visible, in our antiracist and anti-oppression work, the relationships, circles, and groups that defy the idea that nothing will ever change. She writes, "The small circles of love we have managed to form in our individual lives represent a concrete realistic reminder that *beloved community* is not a dream, that it already exists for those of us who have done the work of educating ourselves for critical consciousness in ways that enabled a letting go of white supremacist assumptions and values" (264). She continues, "We need to share not only what we have experienced but the conditions of change that make such an experience possible. The interracial circle of love that I know can happen because each individual present in it has made his or her own commitment to living an anti-racist life and to furthering the struggle to end white supremacy will become a reality for everyone only if those of us who have created these communities share how they emerge in our lives and the strategies we use to sustain them" (271–72). All of this reminded me about my own circles of love and solidarity that I continue to build in my own life and the importance of sharing my own life and process with students who are also yearning, as I have always yearned, for a remade world with love and justice at its center. To me, women's studies provides a powerful context to explore, practice, and build such relationships and communities.

Cultivating Accountability as a Practice of Solidarity

I believe that cultivating a practice of accountability in women's studies programs, classes, and organizations will allow us to build and expand communities of resistance through coalition and alliance building. I choose the term "cultivate" because it implies that accountability is an ongoing process, rather than a destination. Cultivating accountability means intentionally creating a culture within our groups, programs, and organizations where we make it a practice to recognize our implication in systems of oppression and privilege. I am interested in helping to create classrooms and communities where we can challenge and be challenged on the ways in which our ideas, actions, scholarship, and/or activism might reproduce hierarchy. Rather than setting up a structure of innocence and guilt to be met with approval or punishment, we might find more ways to cultivate a sense of community accountability for the perpetuation of oppression and the conditions for transformation.

I continue to struggle with how to engage colleagues, students, and activists in ways that promote individual and cultural change, and that do not simply create distance between us. Although my own anger, frustration, and sense of indignation at others often feels right, it often doesn't translate into the expansion of a just community. I've definitely felt alienated and disillusioned by the level of hostility and self-righteousness that has often attended discussions around these issues. Sometimes we are so hard on one another. We are not going to build a movement based only in negative and self-righteous judgment. I am not suggesting that the problem is anger at oppressive actions and structures. By cultivating accountability for our own participation, and by seeing ourselves implicated in these structures and acting from this knowledge, however, we might be more humble in our approach to one another in the process of change. My hope in the coming years is to collaborate with others in creating more communities of resistance that hold its members accountable in ways that are compassionate and that have faith that we are all capable to change and transformation.

Solidarity, then, is as connected to our political alliances, choices, and actions and accountability for their consequences, as it is to our identities and locations. Again, Morales (1998) writes, "Solidarity is not a matter of altruism. Solidarity comes from the inability to tolerate the affront to our own integrity of passive or active collaboration in the oppression of others, and from the deep recognition of our most expansive self-interest. From the recognition that, like it or not, our liberation is bound up with that of every other being on the planet, and that politically, spiritually, in our heart of hearts we know anything else is unaffordable" (125).

Notes

I want to thank Francesca Royster and Lourdes Torres for their careful reading, their suggestions, and the abundance of support they provided me in writing this essay! I am blessed to have them as part of my beloved community.

1. I take this phrase from bell hooks.

References

Ahmed, L. (1993). *Women and gender in Islam: Historical roots of a modern debate*. New Haven, CT: Yale University Press.

Alexander, M.J. (2005). *Pedagogies of crossing*. Durham, NC: Duke University Press.

Allison, D. (1994). *Skin: Talking about sex, class, and literature*. Ann Arbor, MI: Firebrand Books.

Bannerji, Himani. (1997). "Mary Wollstonecraft, feminism and humanism: A spectrum of reading." In E.J. Yeo (Ed.), *Mary Wollstonecraft and 200 years of feminism*. New York: Rivers Oram Press, 222–242.

Bell, D. (1992). *Faces at the bottom of the well*. New York: Basic Books.

———. (1999). "The power of narrative." *Legal Studies Forum*, 23(3). http://tarlton.law.utexas.edu/lpop/etext/lsf/bell23.htm accessed on December 20, 2007. Reprinted by permission of Legal Studies Forum.

Carrillo Rowe, A. (2005). "Be longing: Toward a feminist politics of relation." *NWSA Journal* 17(2): 15–46.

Carrillo Rowe, A. and S. Malhotra. (2006). "(Un)hinging whiteness: Pedagogies and paradoxes of Whiteness." *International and Intercultural Communication Annual* 29: 166–192.

Chaudhuri, N. and M. Strobel. (1992). *Western women and imperialism: Complicity and resistance*. Bloomington, IN: Indiana University Press.

Davis, A. (1981). *Women, race, and class*. New York: Random House.

Fellows, M.L. and S. Razack. (1997–1998). "The race to innocence: Confronting hierarchical relations among women." *Journal of Gender, Race, and Justice*, 1: 335–352.

Hanh, T.N. (1987). *Being peace*. Berkeley, CA: Parallax Press.

Hobgood, E. (2000). *Dismantling privilege: An ethics of accountability*. Cleveland, OH: Pilgrim Press.

hooks, b. (1981). *Ain't I a woman?* Boston, MA: South End Press.

————. (1996). *Killing rage, ending racism.* New York: Owl Books.

Hurtado, A. (1996). *The color of privilege: Three blasphemies on race and feminism.* Ann Arbor, MI: University of Michigan Press.

Joseph, G. and J. Lewis. (1981). *Common differences: Conflicts in black and white feminist perspectives.* Garden City: Anchor-Doubleday.

Kaufman, C. (2001). "A user's guide to white privilege." *Radical Philosophy Review,* 4(1/2): 30–39. http://www.cwsworkshop.org/pdfs/WIWP2/8Users_Guide_to_White_Priviledge.pdf accessed on December 10, 2007.

Keating, C. (2005). "Building coalitional consciousness." *NWSA Journal,* 17(2): 86–103.

Kimmel, M. and A. Ferber, Eds. (2003). *Privilege: A reader.* New York: Westview Press.

Koyama, E. (2000). "Akron not resolved yet: Why I made a ruckus at NWSA 2000." http://eminism.org/readings/nwsa2000-racism.html accessed on January 6, 2008.

————. (2001). "Transfeminist manifesto." http://eminism.org/readings/pdf-rdg/tfmanifesto.pdf accessed on January 6, 2008.

Lugones, M. (2003). *Pilgramages/peregrinajes: Theorizing coalition against multiple oppressions.* Latham, MD: Rowman and Littlefield.

McIntosh, P. (1988). *White privilege and male privilege: A personal account of coming to see correspondences through work in Women's Studies.* Wellesley, MA: Wellesley Center for Women.

Mohanty, C.T. (1991). "'Under Western eyes': Feminist scholarship and colonial discourses." In C.T. Mohanty, A. Russo, and L. Torres (Eds.), *Third world women and feminist perspectives.* Bloomington, IN: Indiana University Press, 51–80.

————. (2005). "'Under Western eyes' revisited: Feminist solidarity through anticapitalist struggles." In E.L. Kennedy and A. Beins (Eds.), *Women's Studies for the future: Foundations, interrogations, politics.* New Brunswick, NJ: Rutgers University Press, 72–96.

Moraga, C. and G. Anzaldúa (Eds.) (1981). *This bridge called my back: Writings by and about radical women of color.* Watertown, MA: Persephone Press.

Morales, A.L. (1998). *Medicine stories: History, culture and the politics of integrity.* Cambridge, MA: South End Press.

Newman, L.M. (1999). *White women's rights: The racial origins of feminism in the United States.* New York: Oxford University Press.

Razack, S. (1998). *Looking white people in the eye: Gender, race, and culture in courtrooms and classrooms.* Toronto: University of Toronto Press.

Reagon, B.J. (1998). "Coalition politics: Turning the century." In A. Phillips (Ed.), *Feminism and politics.* New York: Oxford University Press, 242–253.

Rich, A. (1986). Virginia 1906. *Your native land, your life.* New York: Norton.

Roman, L. (1993). "White is a color! White defensiveness, postmodernism, and anti-racist pedagogy." In C. McCarthy and W. Chrichlow (Eds.), *Race, identity, and representation in education.* New York: Routledge, 71–88.

Russo, A. 2005. "The taskforce against white supremacy—A herstory in the making." http://www.nwsa.org/communities/awstfhistory.php accessed on December 12, 2007.

Sandoval, C. (2000). *Methodology of the oppressed*. Minneapolis: University of Minnesota Press.

Segrest, M. (1994). *Memoir of a race traitor*. Boston, MA: South End Press.

Smith, B. (1984). "Between a rock and a hard place." In E. Bulkin, M.B. Pratt, and B. Smith (Eds.), *Yours in struggle*. Brooklyn, NY: Long Haul Press, 130–153.

Ware, V. (1991). *Beyond the pale: White women, racism, and history*. London: Verso.

8

My Life and Work in Women's Studies (Now Gender Studies)

Judith Lorber

When I started teaching in 1970 at Fordham University and then at Brooklyn College in their Sociology Departments, there was no Women's Studies. In the 25 years I taught undergraduate and graduate students, I was involved in setting up Women's Studies programs and devising the curriculum for required courses and electives in Women's Studies and on gender in sociology. I was the first coordinator of the Women's Studies Certificate Program at the City University of New York (CUNY) Graduate Center. From that time to the present, I have been doing research on gender issues, and writing books and articles on the subject. I was one of the "founding mothers" of Sociologists for Women in Society (SWS), a professional activist and mentoring organization that began in 1972 and now has over 1,000 members. In 1987, I became the founding editor of SWS's official journal, *Gender & Society,* which is one of the premier academic journals in social science Gender Studies. What I have, then, is an insider's view into the developments, shifts, and conflicts in Women's Studies.

Teaching

One of my first courses was a two-credit class—Courtship and Marriage (a.k.a "dating and mating"). It was held in a large lecture hall, and students sitting in the back often spent the class hour making out. But when I started talking about feminist issues—such as who has power in the family, who controls the money, why women change their names on marriage—they started listening. That May 1970, the students at Brooklyn College and many other campuses in the United States went out on strike over the killing of four students protesting the Vietnam War on the Kent State campus in Ohio. The Brooklyn College faculty supported the strike, and in order to ensure that students would be able to get their credits and graduate on time, we held informal classes in the student union. The students later told me that my class was one of the most eye-opening they had ever had.

My goal was always to open students' eyes to the ways their worlds were organized by gender. As I continued teaching in the 1970s and 1980s, the names of the classes I gave in Sociology reflected my changing thinking about the goals and purposes of the Women's Studies/ feminist perspective that imbued these classes. First, they were called Male and Female in American Society, then the Sociology of Sex Roles. The focus was individual identity, relationships between women and men, family roles, and discrimination against women in the workforce. Then I developed a curriculum for Sociology of Gender, which was a social constructionist analysis of gender as a status as well as an identity. I examined processes of "doing gender" in the socialization of children, in the family, and in the paid workforce, and went on to discuss issues of power, politics, and global gender inequality.

In the Women's Studies programs at Brooklyn College and at the CUNY Graduate Center, I helped develop and taught various versions of introductory courses for beginning students. The year-long introductory class in the Women's Studies Certificate Program at the Graduate Center was divided between a Social Science semester and a Humanities semester. The Social Science course focused on the gender structure of contemporary society. The Humanities course focused on women's history and women's cultural productions. I also team-taught the capstone thesis and dissertation workshop for completing students in the Graduate Center's Women's Studies Certificate Program. These classes were interdisciplinary in that they were team-taught by Social Science and Humanities faculty.

In the 1970s and 1980s the undergraduate students in the Women's Studies and Sociology of Gender classes that I taught were very eager to learn about feminism, and very open to feminist ideas and perspectives. The students were mostly women, but I always had men in my classes, and some I thought might be particularly resistant (Orthodox Jews) were intrigued by gender theory and even recommended my course to friends. In contrast, more conservative women were resistant to these ideas and some even dropped a course rather than have to confront ways of thinking that challenged their values. But by the time I retired from teaching in 1995, after 25 years, students were taking Women's Studies and Sociology of Gender courses to fill a requirement or because they thought they would be easy courses. There were always one or two feminist women who were eager to learn and engage the issues, but the majority of the students were disengaged.

Structure and Focus of Women's Studies

At my own university, both on the college campuses and at the Graduate Center, the Women's Studies programs are interdisciplinary and are taught by faculty whose appointments and specialty areas are in standard academic departments. The problem with this structure is that faculty are "on loan" to the Women's Studies programs for required courses and electives mounted by those programs, or their courses are cross-listed if they have a sufficient Women's Studies component. It is a compromise between mainstreaming or including Women's Studies into the general curriculum and developing Women's Studies as an interdisciplinary area. In my university, I did not see the latter development to any great extent.

Because I am a social constructionist that sees gender as a building block in a social order that is organized by the division of people into men and women, I soon became dissatisfied with a focus just on women. In my mind, what we needed to study and analyze and try to change was the social relationships among women and men of different groups, and the inequality that results from legal and informal forms of gender discrimination and how these are intertwined with social class, racial ethnic, and other forms of discrimination. I moved from blaming the patriarchy and valorizing women to trying to understand how gendered societies are constructed through gendering processes and maintained through gendering organizations. I recognize the power of patriarchal privilege, but also think that just as not all women are oppressed, not all men are oppressors. Some, as we all know, are oppressed themselves. As a result of this shift in my thinking, I began to use the concept of gender exclusively, and to say that my field is Gender Studies.

When Gender Studies was first proposed, there was a fear that women would once again become invisible, since Gender Studies recognizes that men as well as women have gender. In my work, I used the concept of gender as a social status in a social order based on gender divisions. Thus, I think it is important to study the structure and institutionalization of those divisions throughout the social order (e.g., work, family, politics, education, military, state regimes are all organized by gender). Gender Studies most notably deconstructs the social processes that create and maintain that structure and institutionalization—what we call "doing gender." Everyone does gender, but we want to know how women and men of different social groups (by ethnic identity, religion, nationality, and social class) do gender and thus how ethnicities, religion, nationality, and social class are all gendered. Gender Studies allows us to study these intersections for different societies. Gender Studies also allows us to analyze the intersections of gender and the body, gender and sexuality, gender and personality.

For me, Gender Studies encompasses but goes beyond Women's Studies in that it examines the whole structure and process of the gendered social order we create, maintain, and are gendered within.

Research and Writing

My teaching was shaped by the way the field developed; as books and articles were written I used them—mostly readers and monographs rather than textbooks. As time went on, I produced my own articles and books, many of which I used in my classes. *Paradoxes of Gender*, published in 1994, is still being used as a text throughout the world. Since retiring from teaching, I have written three editions of *Gender Inequality: Feminist Theories and Politics* (3rd edition, 2005b), which have been well received as texts in Women's Studies courses. A fourth edition will be published in 2009. With Lisa Jean Moore, I published the second edition of *Gender and the Social Construction of Illness* (2002) and *Gendered Bodies: Feminist Perspectives* (2007), which addresses the burgeoning interest in the body current in Women's Studies. My other books in gender studies are *Women Physicians: Careers, Status, and Power* (1984), based on comparisons of the careers of women and men physicians, and *Breaking the Bowls: Degendering and Feminist Change* (2005a), the culmination of my thinking about how we could "do degendering." By degendering, I mean challenging the legal rigidity of binary gender statuses, their constant use in the allocation of family work and paid jobs, and the embedded notion of men's entitlement to women's services and sexuality.

One of the issues I have addressed throughout my work is the multiplicity of women's and men's standpoints and identities. The concept of *standpoint*, or perspective from one's social location, is an important one in Women's Studies. In the 1970s, there were many in Women's Studies who argued that women's viewpoint must be privileged in order to counteract the dominance of men's perspectives. They argued that men set the agendas for scientific research, shape the content of higher education, choose the symbols that permeate cultural productions, and decide political priorities. Women's Studies therefore insisted that women's perspective, women's way of seeing the world, be privileged in the production of knowledge and culture, in setting political agendas, and in doing research.

But is all women's experience the same? Donna Haraway claims that all knowledge is partial, dependent on social location, situated somewhere within limits and contradictions. Thus, all views are views from somewhere (1988). Sandra Harding, who did the groundbreaking work on bringing women's standpoint into science, raised the question of fractured standpoints in 1986 and now argues for the multicultural aspects of science. In postcolonial feminist science studies, Harding (1998) notes that where Northern feminists critique the exploitative and dominating concepts and practices of technologically based science, Southern feminists look at the lives of African, Chicana, Asian, Indian, Native American, Aboriginal, Maori, and other women. They critique the negative effects of technologically based science on economic restructuring and imperialist expansion.

Patricia Hill Collins (1999) is critical of studies of Western women in science that ignore their other social characteristics. She says, "If the absence of women is critical in the production of scientific knowledge, then the absence of racial, ethnic, and social class diversity among women who critique science certainly must have an impact on the knowledge produced. Whether intentional or not, feminist scholarship on scientific knowledge seems wedded to the experiences of White, Western, and economically privileged women" (267). What is needed, she says, is *intersectional analysis.* "This approach means choosing a concrete topic that is already the subject of investigation and trying to find the combined effects of race, class, gender, sexuality, and nation, where before only one or two interpretive categories were used" (278).

I have felt that a critical intersectional perspective was crucial to advance Women's Studies. It would break up the concepts of gender as two and only two "opposites" and help to advance what I see as a potentially revolutionary concept—*degendering.*

Women's Studies and Feminist Research

One of the contested areas in Women's Studies has been doing research on feminist activist groups whose goals the researcher would like to support. First, there is the problem of the *insider* and the *outsider*. If you are a member of the group, you don't have to study it—just relate your own experiences. And for some feminists, that is what feminist research should be. But there may be a question of divided loyalty if the goals of objectivity in research and data analysis conflict with the goals of loyalty to the group and support of its work.

If you are not a member of the group, but an outsider, there is the problem of trust and betrayal of the group's confidences. Even if you were honest and above board about your role as a researcher, and you are careful to preserve anonymity, people may be hurt when you publish your findings—especially if you are critical. In asking whether one can be a reflexive participant observer and a true feminist, Judith Stacey honestly appraised the advantages and disadvantages of ethnography and her own use and response to intensive interviewing and involvement with her respondents' lives. She discussed the "delusion of alliance" and the "delusion of separateness." She says:

> ...[E]thnographic method appears to (and often does) place the researcher and her informants in a collaborative, reciprocal quest for understanding, but the research product is ultimately that of the researcher, however modified or influenced by informants. With very rare exceptions, it is the researcher who narrates, who "authors" the ethnography. In the last instance, an ethnography is a written document structured primarily by a researcher's purposes, offering a researcher's interpretations, registered in a researcher's voice. (1988, 23)

To do any kind of research—sociology or anthropology or history or literary or art criticism—the scholar has to bring to bear a somewhat distanced stance, has to be uncomfortable with the contradictions in data because they are likely to be crucially informative, and has to be able to challenge respondents' voices with voices from other worlds. Sometimes you have to be able to look at the familiar world as if you came from another planet. As Dorothy Smith says, we cannot take the everyday world for granted but must see it "as problematic, where the everyday world is taken to be various and differentiated matrices of experience—the place from which the consciousness of the knower begins..." (1990, 173).

When Women's Studies researchers construct the patterns of social reality from everyday experiences of subjects, they do it from the standpoint of their own social realities. Even if the researcher and the subject are the same gender, they are not likely to come from the same social location. Even if they did, the social researcher still needs a bifurcated consciousness that can bring to bear a somewhat abstracted larger social reality (the social relations of capitalism, for instance) on the patterns and experience of the everyday world.

Feminist Perspective and Radical Thinking

The relationship between feminism and Women's (or Gender) Studies has been contested, but to me, there is no question about the feminist perspective as the guiding spirit of the field—and of course, my own work. As I see it, the feminist perspective is challenging received wisdom by asking why we do gender, what is the outcome, and should we continue in the same way. A feminist viewpoint often thinks what some may claim is impossible—in my work, thinking about a world without gender, without a constant division of everything into male and female, and without the privilege and centrality of men and masculinity.

Feminism has moved from a focus on women's oppression to recognition of the nuances of the intersectionality of gender, social class, racial ethnic, and other statuses that create the conditions of complex inequality. Another way that feminism has challenged conventional ways of thinking is by going beyond the binaries and using concepts of multiple sexes, sexualities, masculinities, femininities, and genders. Data that undermine the supposed natural dichotomies on which the social orders of most modern societies are still based could radically alter political discourses that valorize biological causes, essential heterosexuality, and traditional gender roles in families and workplaces.

In my research on women and men physicians, I found that the behavior of men and women doctors sometimes reflects their professional status and sometimes their gender, and it is important to look at both aspects to understand their relationships with patients (Lorber, 1984).

Sociologists for Women in Society

The main Women's Studies/feminist network that I have been involved in is Sociologists for Women in Society (SWS), both nationally and as organizer and president of the New York Metropolitan Area Chapter in 1971–1972. I was secretary of national SWS in 1974–1977, and president in 1980–1982. I also helped form and was chair of the International Committee from 1993–1998, when I worked toward expanding SWS's presence at the United Nations. I was founding editor of its official journal, *Gender & Society*, in 1986–1990.

SWS both shaped and was shaped by feminist social science as focus of research and writing and by feminist activism. Its early mission as a professional organization was to expand the presence of women in American Sociological Association (ASA), to mentor and support women faculty and students, and to mainstream women's and gender studies on campuses and in curricula. It has been highly successful in all three endeavors. Through *Gender & Society*, SWS provided very important, and widely cited, theoretical articles that have become Women's Studies classics.

Today, SWS has many men members, two of whom were SWS feminist lecturers, and it includes political and social action in its mission. From its Web site (http://socwomen.org),

Sociologists for Women in Society is an international organization of social scientists—students, faculty, practitioners, and researchers—working together to improve the position of women within sociology and within society in general. Our goals are to:

- promote and disseminate research about women—on race and gender inequity in pay, the struggles of women of color and working class women, on the feminization of poverty and the effects of welfare reform, on issues of health, sexuality, invisibility, violence, and sexual harassment.
- educate our sociological colleagues as well as the general public and government officials about the results and implications of research on women.
- help women to establish careers as sociologists through professional development workshops at national and regional meetings, a mentoring program, supporting a minority fellowship, and providing scholarships to support research.
- offer help and mutual support to women and men who share our concerns through an information network and friendship.
- take political action to improve women's lives by preparing resource papers for use by policy makers, supporting feminist candidates for public office, consulting with feminist groups on research and data analysis, lobbying and testifying at legislative sessions.

Some of the activist accomplishments listed on the Web site are: helping to support an ASA minority scholar, providing financial support to women sociologists waging legal battles against discrimination, supporting research on breast cancer with the Barbara Rosenblum Cancer Dissertation Award, recognizing people and local organizations with the Pauline Bart Feminist Activism Award, and working with the UN Committee on the Status of Women and other UN groups to promote women's rights internationally.

Future of Women's (Gender) Studies[1]

In my view, the future of Women's Studies lies in recognizing the variety and density of multiple identities and moving away from an exclusive focus on women. Its strength is its sensitivity to multicultural perspectives and not imposing Western values on research questions and data analysis. I do not think that the conflict between research and activism has been resolved because of the difficulties of maintaining a critical distance while doing research on feminist groups.

Women's Studies is already morphing into Gender Studies, as many new programs call themselves. (And many older programs are changing their names to Women's and Gender Studies.) Gender Studies often incorporates feminist studies of men and sexuality studies as well. The strength of Gender Studies is that it recognizes the multiplicity of genders, sexes, and sexualities. The problem here is that we need categories for comparison, even while we are critically deconstructing them. We know that the content and dividing lines for genders, sexes, and sexualities are fluid, intertwined, and cross-cut by other major social statuses, but are in the process of figuring out how to research these complexities.

In the past, most Women's Studies research designs assumed that each person has one sex, one sexuality, and one gender, which are congruent with each other and fixed for life. Research variables were "sex," polarized as "females" and "males;" "sexuality," polarized as "homosexuals" and "heterosexuals;" and "gender," polarized as "women" and "men." But these vary, and for accurate data, we need the variations. How do we include masculine women, feminine men, bisexual women and men, intersexuals, transsexuals, cross-dressers, and their partners? We also need to compare women and men across different racial ethnic groups, social classes, religions, nationalities, residencies, and occupations. In some research, we need to compare the women and men within these groups.

The main question is, who is being compared to whom? Why? What do we want to find out? The goal of our research design needs to be clear before we can decide on our comparison categories. The choice of categories is a feminist political issue because using the conventional categories without question implies that the "normal" (e.g., heterosexuality, boys' masculinity) does not have to be explained as the result of processes of socialization and social control, but is a "natural" phenomenon.

Deconstructing sex, sexuality, and gender reveals many possible categories embedded in social experiences and social practices, as does the deconstruction of race and class. Multiple categories disturb the neat polarity of familiar opposites that assume one dominant and one subordinate group, one normal and one deviant identity, one hegemonic status and one "other."

In political activism, can we recognize gender diversity and also continue to rally around the flag of womanhood? Judith Butler says, "Surely, it must be possible both to use the term, to use it tactically...and also to subject the term to a critique which interrogates the exclusionary operations and differential power-relations that construct and delimit feminist invocations of 'women'" (1993, 29). Is it possible to conceptualize Woman as a stable category and to use the experiences of women with varying other important identities at the same time? There may be a common core to women's experiences, but feminist politics cannot ignore the input from social statuses that may be as important as gender. Therefore, it is not enough to take political action using only a woman's point of view; feminist politics has to include the experiences of women of different social classes, educational levels, racial, ethnic, religious, and groups, marital and parental statuses, sexual orientations, ages, and degrees of able bodiedness. These intersecting group memberships and multiple identities are not a weakness in the feminist movement, but a strength. If they become a basis for national and international coalitions, cross-cutting group memberships and multiple statuses can be a powerful feminist weapon.

For myself, I will continue to work in Gender Studies and SWS, and bring my feminism to all my research, writing, and politics.

Note

1. Adapted from Lorber, 2005b.

References

Butler, J. (1993). *Bodies that matter: On the discursive limits of "sex."* New York: Routledge.

Collins, P.H. (1999). "Moving beyond gender: Intersectionality and scientific knowledge." In M. M. Ferree, J. Lorber, and B. B. Hess (Eds.), *Revisioning gender.* Thousand Oaks, CA: Sage.

Haraway, D. (1988). "Situated knowledges: The science question in feminism and the privilege of partial perspective." *Feminist Studies,* 14: 575–99.

Harding, S. (1998). *Is science Multicultural? Postcolonialisms, feminisms, and epistemologies.* Bloomington: Indiana University Press.

Lorber, J. (1984). *Women physicians: Careers, status, and power.* New York: Tavistock.

———. (1994). *Paradoxes of gender.* New Haven, CT: Yale University Press.

Lorber, J. and L.J. Moore. (2002). *Gender and the social construction of illness.* 2nd edition. Lanham, MD: Rowman and Littlefield.

———. (2005a). *Breaking the bowls: Degendering and feminist change.* New York: W.W. Norton.

———. (2005b). *Gender inequality: Feminist theories and politics.* 3rd edition. New York: Oxford University Press.

———. (2007). *Gendered bodies: Feminist perspectives.* New York: Oxford University Press.

Smith, D.E. (1990). *Texts, facts, and femininity: Exploring the relations of ruling.* New York: Routledge.

Stacey, J. (1988). "Can there be a feminist ethnography?" *Women's Studies International Forum,* 11: 21–27.

9

Continuity and Change in Women's Studies Programs: One Step Forward, Two Steps Backward

C. Alejandra Elenes

Introduction

The contributions of women of color to feminist scholarship are well documented (Anzaldúa, 2007; Collins, 2000; Guy-Sheftal and Hammond, 2005; Mohanty, 2003; Sandoval, 2000), demonstrating their significance to the evolution of women's studies.[1] As a field women's studies has progressed from exclusionary practices and scholarship that focused on the experiences and perspectives of white women (Baca Zinn et al., 1986; Sandoval, 1990) to a more inclusive model (hooks, 2000). Indeed by the end of the twentieth century, women's studies adopted the early demands (e.g., Davis, 1981) for the implementation of an intersectionality approach to the study of women and gender. This is not to say that women's studies has become a unified area of study; there are political, economic, and social differences among various groups of women that can lead to different goals (Brown and Chávez-García, 2005). How can we account for those different goals and agendas in women's studies? What type of institutional practices can facilitate or hinder the development of women's studies with wide-ranging perspectives?

In this essay, I want to answer these questions by examining my experiences in women's studies as a Chicana/Mexicana working in a program that has struggled to gain institutional respect. In doing so, I will analyze the progression of the Women's Studies Program I have been a faculty member for 16 years. I conduct this analysis by examining the development of the Program's vision, a discussion of its structural location within the university campus, and the faculty's struggle to survive in a constantly changing academic environment. I believe that the evolution of the Women's Studies Program at Arizona State University (ASU) at the West campus can offer a glimpse to the advances and limitations of women's studies in the United States at the end of the twentieth century and beginning of the twenty-first century. My intentions are not to offer a genealogy of women's studies in general or even of Chicanas and women of color within women's studies; but to reflect on the particular history of a program that was developed in the 1990s and in doing so was able from the beginning to organize itself through intersectional analyses of race, class, gender, and sexuality both in United States and transnational contexts.

Institutional Context

Arizona State University (ASU) is composed by four campuses: Tempe, Downtown, Polytechnic, and West. The Tempe campus is the original and largest campus, which throughout the 1990s and early 2000s was called ASU's Main campus. ASU's mission is described as the New American University, and proposes to develop a new standard in higher education by pursuing excellence, access, and social and economic impact. The four campuses are lead by a president, and since 2007 ASU was redesigned to a college-centric model, where deans report directly to a university provost, regardless of the campus where they are located. The central administration is on the Tempe campus. In its 20-year history, the West campus of ASU has gone through a series of changes that have affected its development in both positive and negative ways. It is within a context of multiple structural changes within the University and the campus, that I will discuss the evolution of the Women's Studies Program at ASU's West campus.

The Early Years

The Arizona Legislature authorized the creation of a new Arizona State University campus on the West side of the Phoenix metropolitan area in 1984 as an upper-division university. The current campus was inaugurated in 1989. The presence and existence of women's studies at ASU West dates back to 1989. It was designed as a freestanding entity with faculty who would have full-time appointments and tenure home in Women's Studies. In its 19-year history, the program has been moved from different structural (and in one case physical) locations in the campus organization. When the program was first approved by the Arizona Board of Regents (ABOR) it was housed in the College of Arts and Sciences. Sometime around the late 1980s and early 1990s, the Women's Resource Center was also approved with a director's line. When a chair whose specialty was in Social Work was hired, the program was moved to a division called Inter Unit Programs. Women's Studies was the only academic program in such a division. In 1994, Inter Unit Programs was renamed the Division of Collaborative Programs, and Women's Studies continued to be part of that unit. From its inception, the Women's Studies Program at ASU West differed from other programs and departments due to its structural location. This structural location has placed the program in an awkward position: it has had some autonomy, but not enough support.

The needs of the program have not been the foundation for all of these moves to different organizational structures. Rather, they have been instigated from political expediency on the part of university administrators. When the inaugural coordinator/chair[2] was hired (around the late 1980s or early 1990s), Women's Studies was moved to Inter Unit Programs. When I came to my campus interview in 1992 and realized that Women's Studies was housed in a unit outside traditional schools and colleges (e.g., Arts and Sciences), I thought this was a strength because it seemed to give Women's Studies an autonomous space. However, once I arrived at campus and was aware that we were the only academic program in such a unit, it became clear to me that the structure was more akin to marginalization or ghettoization. Women's Studies' position within the organizational structure was very tenuous; it had a coordinator/chair but no dean to report. Reporting lines were unclear, it was very likely that the chair/coordinator reported to the associate vice provost for academic affairs.

As I arrived to campus in the fall of 1992, I was informed that the provost of the campus, who served as its chief academic officer, had resigned. The Women's Studies coordinator who had hired me was promoted to the vice provost for academic affairs position, and in a somewhat ironic move, the wife of the previous provost, who did not have a faculty appointment, was named coordinator. She did not have any power, thus we did not have an advocate for the program, even though a top administrator of the campus had her faculty appointment in Women's Studies. Positioning Women's Studies in a unit by itself was done for the advantage of an administrator who wanted to ensure a tenure home outside of Arts and Sciences, and not for the benefit of the program. In a campus and program that values interdisciplinary scholarship it should have not mattered what the area of expertise of the chair was. A quality scholarly record commensurate with the level necessary for the appointment should have sufficed.

In 1992 when I commenced my academic career in Women's Studies at ASU West, I was not the only assistant professor hired. Three of us arrived at the same time, and all three identified as Latinas. Although for some this might seem as a homogenous group of faculty, nothing is farther from the truth. We came at different stages in our academic career, indeed I was the only recent graduate (in reality I was ABD ["all but dissertation"]). Demographically, we were diverse too: one was Puerto Rican, one was Argentinean, and I identify as a Chicana/Mexicana. Already on board was an African scholar. All of the faculty were tenure track, and we were all in need of mentoring. The coordinator who was the wife of the previous provost stayed for only one semester; a senior faculty member from education without any experience in women's studies was appointed interim coordinator. She neither had the skills nor the desire to mentor junior faculty members in feminist scholarship.

At the time I joined the faculty ranks, I was coming from the experience of a graduate student activist. Actually, my arrival into women's studies was via Chicano studies. There are parallels in the development of alternative and interdisciplinary academic spaces in higher education such as women's studies, Chicano/a studies, African American studies, American Indian studies, Asian/Pacific studies, and GLBT/Queer studies, as they all are the result of years of activism and struggle. Moreover, these interdisciplinary spaces share in common a commitment to the development of alternative epistemologies and methodologies that seek to decenter the hegemony of Western thought and subjectivity. I got involved in higher education politics first as a student activist demanding the development of Chicano Studies at the University of Wisconsin-Madison (see Casanova, 2001).[3] Because I first engaged with the development of Chicano studies, my participation in women's studies has always centralized and integrated as the analysis of race, ethnicity, gender, and class.

As many women who got their graduate degree in 1980s, I do not have an M.A. or Ph.D. in women's studies, but I do have an interdisciplinary background. As a graduate student at Wisconsin, I assembled an academic program of study that would help me develop the academic foundation in feminist scholarship, particularly those developed by Chicanas and other women of color. This meant finding the few graduate courses offered in Chicano studies, and combining them with courses in education, my chosen graduate field. I was fortunate that I found courses in education that examined race, ethnicity, gender, class, and sexuality.

I also started to read some of the foundational texts written by women of color and Chicana feminist such as Cherrie Moraga and Gloria Anzladúa's *This Bridge Called My Back* (1983), Gloria Anzaldúa's *Borderlands/La Frontera*, bell hooks, *Feminist Theory: From Margin to Center* (1984), Patricia Hill Collins *Black Feminist Thought* (2000), Paula Gunn Allen *The Sacred Hoop* (1986), and Asian Women United of California, *Making Waves* (1989). I began to develop expertise on women of color feminisms, which eventually led to my formal entrance to, and eventual career in women's studies. In 1989 when the Women's Studies Research Center, which was associated with the Women's Studies Program at the University of Wisconsin-Madison, was searching for a graduate assistant to coordinate a curriculum transformation project to integrate the "new" scholarship on women of color called the Women of Color in the Curriculum (WOCC) project. I was strongly encouraged to apply by Chicano Studies and I got the job. Working on the WOCC project was my official entry into women's studies.[4]

Arriving in Arizona as an ABD was a mistake, I felt at a disadvantage in relation to my peers. The two other Latina colleagues who were hired at the same time as I was had some experience as assistant professors, and had already started developing a research agenda and publication record. I was writing my dissertation at the same time that I fulfilled all the duties of a faculty member I taught a full load and provided service to the program, campus, and university. The four tenure-track faculty were in different stages of their professional development. But the four of us were untenured and in need of guidance. Moreover, we represented different viewpoints, geographical locations (both national and international), and personalities. We had to negotiate these differences at the same time that we were trying to understanding a complex structural location within the university, and get tenure.

The administrative instability at ASU West has not only plagued Women's Studies but also the campus as a whole. During these early years, we had a provost who had been appointed for only three years (he replaced the previous provost who resigned in spring of 1992). Stability at the top-level administration of the campus was not achieved until 1996, when a provost was hired served until 2004. As a new campus developing its academic programs, there was a large amount of work that needed to be done. The service load of the faculty was excessive, especially for minority faculty. In addition, the majority of the faculty in the campus as whole was also untenured. The difficulties of successfully negotiating tenure while building a campus were particularly taxing for the Women's Studies faculty. Because we were located in a unit that operated like a college, we were obligated to have representatives in all of the campus committees as if we were a college. On the one hand, this situation granted Women's Studies visibility and voice in campus decision-making; on the other hand, it created an excessive amount of service for untenured faculty. That is, in the "regular" colleges the faculty would vote for a representative on a particular committee among various departments; Women's Studies only had four faculty members to serve on a whole array of campus committees.

The Women's Studies faculty served on several committees while we had to teach a full load of courses and publish. This was extremely difficult and arduous, especially because it was done without proper guidance and support. This also placed Women's Studies in a complicated position vis-à-vis the rest of the campus. At a first glance, it seemed that Women's Studies was at the nucleus of the governance structure of the campus as its faculty were represented in almost every conceivable committee. But some times too much visibility can hurt and be as problematic as marginalization. The presence of untenured faculty in some committees was not always welcome, and put some of the faculty in vulnerable positions and lead to burnout. For example, in my first year as a tenure-track faculty (which was my second year at ASU West) I was asked to serve on the College of Education Dean Search Committee. My efforts to ensure that affirmative action was seriously considered put me in a terrible position. At the end of the search, when the internal candidate was selected, I realized that I had not only dedicated too much time in this process, but that I had made some enemies.

Centering Women of Color and Global Perspectives

As the Women's Studies faculty toiled to help build the ASU West campus, we had to work toward the development of our program too. The Women's Studies Program at the Tempe campus precedes the Program at the West Campus. When Women's Studies was approved by the Board of Regents at the West campus it imported[5] the curriculum from Tempe. Over the years, Women's Studies at the West campus transformed the inherited curriculum to align it with our vision and the expertise of the faculty. Women's Studies engendered an ideology in the 1990s that reflected a commitment to feminist scholarship that valued and centered "third world" women feminisms.[6] Consciously the faculty worked to put into practice a feminist curriculum that underscored the centrality of U.S. women of color and "third world" women to the field. Women's studies documents, such as the mission statement, promotion and tenure guidelines, and publicity pamphlet, emphasized this vision by explaining our mission as an U.S. cross-cultural and global perspective to the study of women and gender. Based on this philosophy, in 1994 we developed a curriculum where intersectionality and transnationalism/global standpoints were central. For example, the core curriculum included four courses: Introduction to Women's Studies; Race, Class and Gender; Women, Cultures and Societies (which has a global perspective); and the Pro-Seminar: Theory and Methods in Women's Studies. The rest of the curriculum was divided into three areas of study: Global, Representation, and Cultural Studies. The curriculum also included electives and a related field.

In addition to centralizing intersectionality and transnationalism, the interdisciplinary aspect of the curriculum is shown in the emphasis on areas of concentration rather than in disciplinary study. The curriculum avoids an over emphasis on the "women and…a particular discipline" (e.g., history) approach, although this is not always avoidable or should always be avoided. Perhaps the vision of the program and its emphasis is best exemplified in the way in which the pro-seminar is taught. Instead of emphasizing typical strands in feminist theory: liberal, radical, socialist, multicultural, and postmodern, or the focus by nation (e.g., French feminist theory), the course examines feminist theories, methodologies, and epistemologies from an intersectionality approach of race, class, gender, and sexuality within national and transnational contexts. Special emphasis is placed on historical definitions and debates on the relationship between feminist theory and activism. At the same time, the course examines significant feminist theoretical movements in the twentieth and twenty-first centuries. These theoretical frameworks are studied under the lens of standpoint epistemologies and innovative feminist methodologies. An underlining aspect of this approach to theory is the decentralization of the hegemony of Western thought.

Moreover, the interdisciplinary characteristic of women' studies aligns well with its vision toward social, gender, racial, and sexuality justice. The complexity of women's multiple positions in the world, and the uneven worldwide status of women, makes it imperative to apply interdisciplinary and transdisciplinary epistemologies, methodologies, and theories. A unitary definition of women that does not take into account the major discontinuities among women due to race, class, gender, and sexuality as forms of privilege and oppression simply oversimplifies the study of women. From a global perspective, the study of women from a "first world" imperial standpoint only serves to maintain the hegemony of the West. This perspective is similar to the feminist solidarity model proposed by Chandra Mohanty, which "rather than Western/Third World or North/South or local/global seen as oppositional and incommensurate categories, the One-Third/Two-Thirds World differentiation allows for teaching and learning about points of connection and distance among and between communities of women marginalized and privileged along numerous local and global dimensions" (2005, 87).

This perspective offers feminist pedagogical tools where students are able to understand national and global dynamics of power and how they can fight them. In this way, students can understand their multiple subject positions in relation to forms of domination and resistance. But more importantly, they can use these academic tools to see themselves as agents of change in the expansion of local and global social and gender justice.

The vision that Women's Studies engendered at West could have not occurred without the presence of a diverse faculty. The materiality of difference as shown by the presence of women of color, transnational feminists, working-class, and lesbian/transgender faculty, staff, and students makes a difference. As Gloria Ladson-Billings (2000) writes, "how one views the world is influenced by what knowledge one possesses, and what knowledge one is capable of possessing is influenced deeply by one's worldview" (258). In the case of Women's Studies at the West campus, the material presence of women of color made the difference in the formulation of an innovative women's studies curriculum. This curriculum was hotly debated, discussed, and struggled over. And it was accomplished with little institutional support.

As untenured faculty, we were navigating the tenure process, and many of us understood that we had to concentrate on our publications in order to be successful in achieving tenure. Publishing was difficult (nevertheless accomplished) due to the demands of complying with excessive amount of service. One of the ways a group of faculty—composed mainly of Latinas/os (but not exclusively)—achieved tenure was through the formation of a manuscript group. The manuscript group was originally developed by the vice provost for faculty development, but eventually developed into an organic support network that facilitated the progress toward tenure and gave needed emotional support.

At a personal level, at the time that the manuscript group was formed, I found out how difficult it is to balance family and academia. When I was preparing my fourth year review file in the fall of 1996, I found out I was pregnant. This meant that two years before I was going up for tenure I had my son. At that time, ASU did not have a family leave policy other than the federal government's unpaid policy (in 2006 ASU implemented a paid leave policy). Since my due date was in May, I mistakenly believed that I did not need maternity leave thinking that the summer would be enough time to take off. During the summer after my son was born, I wrote one journal article between feeding the baby, changing diapers, and with little sleep. The fall semester I came back full-time and without a babysitter. My husband and I took care of the baby while we both taught our regular full load. By the beginning of the spring 1998 semester, the baby was going to daycare. I had to figure out on my own how to extend my tenure clock. Women's Studies structural location made it difficult to figure out if there was a maternity leave policy and what were the procedures to extend the tenure clock due to maternity. The chair did not know and I had to send more than a few e-mails (this was the first time it occurred to me to navigate the bureaucracy via e-mail so I would have in writing the responses from administrators) to be able to extend my tenure clock (nine years later, a junior colleague had to go through the same process).

Without the support of the manuscript group my colleagues and I would have not been able to successfully obtain tenure. The fact that the four original tenure-track Women's Studies faculty were "successful" in the tenure process, does not mean it was easy and without "casualties." As overloaded junior faculty, by the time we got to the tenure process many of us were suffering burnout, and had too many emotional scars of years of dealing with institutional instability. In 1998, Women's Studies was moved to the College of Arts and Sciences.

"We're Moving to Arts and Sciences"

The move to Arts and Sciences was not smooth, as faculty in Arts and Sciences saw it as the addition of one more department with which they had to share resources. The reality was that some of our budget went to the college. The constant flux that has plagued Women's Studies placed it in a vulnerable position. When ASU West became separately accredited from the Tempe campus, the divisions and programs were finally named colleges and departments. Before it moved into Arts and Sciences, Women's Studies literally fell through the cracks and was not included in the packet sent to the Board of Regents to authorize the new academic structure and did not become a department. Women's Studies was a program, even though it was identified as a department with a chair (in 1994 a "permanent" chair was hired) and faculty lines.

To make matters worse, the first casualties of years of constant struggle and lack of support materialized. In the spring of 1999, the semester before I would go up for tenure, one of the Latina faculty members who had arrived with me in 1992 announced that she had accepted a position in her disciplinary department at the Tempe campus. A few months later, the other Latina also resigned and left academia altogether. It is easy to dismiss these two resignations as personal and academic decisions that do not have anything to do with the working environment. However, the fact that one chose to move to a department that offered her stability and resources with which the West campus could not compete, and the other suddenly after being granted tenure opted to leave academia altogether, should make us take pause on the consequences of the structural instability.

In addition to the personal loss of colleagues, losing two lines at that particular time was devastating. The campus had stagnated in its growth, partly because it was only an upper division university. The state legislature was not going to allocate any more resources to the campus until it reached the goal of 5,000 FTE (Full Time Equivalent) students. The college that suffered the most was Arts and Sciences. A characteristic of the West campus, especially throughout the 1990s, was that it attracted first-generation, mostly women, students. These are the type of students who view their education in very pragmatic ways and tend not to enroll in many Arts and Sciences majors (with the exception of Psychology and Life Sciences). Without the lower division General Studies courses that are normally offered by Arts and Sciences, it was very difficult to grow. Women's Studies courses had a healthy enrollment that increased every year; but it had a small number of majors (which is very characteristic of women's studies degrees). The two lines vacated by the two resignations were lost. Women's Studies was left with three faculty members, one who served as chair to offer, govern, and maintain the program.

In 2000 the West campus started to offer lower division courses, which help increase the student FTE significantly. By 2002, the campus reached its goal of 5,000 FTE, but there was a shortfall in the state budget due to the events of 9/11, and there were few lines allocated. Women's studies was not prioritized and was not granted permission to hire new faculty until 2004. This occurred when affiliated faculty who served as chairs of other departments in the college pointed out that Women's Studies had a student faculty ratio of 80:1. Yet, the first search approved was for a lecturer, and it was not a new line. The dean reallocated the line of the director of the Women's Resource Center as a lecturer line. The director of the Center was fired in the summer of 2003 and the faculty were not notified. In 2005 and 2006 Women's Studies was given two lines (one each year) and two junior faculty members were hired.

Over the years, our student body has been diverse. Over 50 percent of our students are minorities, and of these 25 percent are Latinas. The relatively low number of majors is a persistent problem that plagues the program. Women's Studies courses carry General Studies requirements; therefore, most of the courses end up as service classes. From a feminist education perspective this is an incredible contribution to the campus. Students in Business, Social Work, and most of the majors in Arts and Sciences take our courses and are exposed to feminist pedagogical principles. Women's Studies serves a high number of students. In this sense, Women's Studies is very productive, and provides an important service to the university and the campus. But the numbers of majors, compared with Psychology, for example, are low. The numbers of majors fluctuate from year to year; the number of majors have ranged from a low of 10[7] to a high of 25 or even 30. For a program with five lines, this number might seem small. But proportionally, Women's Studies at West is at par with the field. Moreover, the University refuses to adequately count double majors, counting only the first major a student declares. This disadvantages Women's Studies because students tend to declare Women's Studies after they have taken courses in the field. Many do not know what women's studies is (or what liberal arts degrees are) and much less what they will be able to do with a degree in women's studies. Students tend to stumble into Women's Studies courses because they have General Studies and sound interesting. Once they take the courses, some find the field exciting and many opt to do a double major.

Women's Studies faculty have recognized that it is necessary to recruit students. Yet, this is a task that is very difficult to accomplish when the faculty are loaded with so many demands on their time. In the first years, the faculty were burdened with building a campus and a program while they were working toward tenure. Then, the number of faculty shrunk, and those left had to concentrate on the day-to-day survival of the program. By 2006–2007 when the program was functioning again in full force, we were starting to work toward the recruitment of students. In the fall of 2007, Women's Studies was able to combine the administrative assistant duties with advising, a model aimed at increasing the number of majors (and early evidence shows that the model is working, as the number of majors is increasing).

I understand why students seek pragmatic majors, as an undergraduate student in the 1970s I did not know what one could do with a liberal arts degree. I opted to major in journalism because it seemed to offer a clear path toward employment after graduation and the courses looked interesting. I remember the first time I heard of women's studies. It was about 1983 (when I already had a college degree and was thinking of graduate school) and I was living in Milwaukee. I was walking on the campus of the University of Wisconsin-Milwaukee, and saw a huge banner that either demanded the development or announced the inauguration of (I cannot remember) a Women's Studies Program. The banner caught my eye, and I was intrigued. What was Women's Studies? Do people actually concentrate on the study of women?

Until then, my educational experience was of a male-centered curriculum where women were either absent or distorted. Initially, I didn't see myself fitting within a women's studies paradigm, but I gravitated toward feminist scholarship. I had internalized the normalizing processes that gave rise to the hegemony of a male-centered curriculum. I had honestly not given much thought to the absence, silence, and marginalization of women in my education. It just seemed so "normal!" The first time, I consciously realized that women had been absent in my education was when I was having a conversation with my roommate (sometime in the early 1980s) as she informed me that she loved to read women's literature. And in one of those aha! moments, it dawned on me that I had never thought of women's writings or scholarship as significantly different from men's writings. And more importantly, I started to realize what this curricular silence meant. I believe that this sudden realization lead to my interests in feminist scholarship. So, that day when I saw the banner announcing or demanding women's studies, I was intrigued and wanted to know more about what this academic space meant. It would take me about four or five years to get officially involved in women's studies. I bring this experience to my courses and in discussion with students I help them understand, or being cognizant of, the fact that the educational system encourages us to think pragmatically and not be socially conscious. Nevertheless, I also let potential students know what some of our graduates have done with their women's studies degree, which includes careers in the public, private, and nonprofit sections. Other paths students have taken are to enroll in advanced and professional degrees.

The Future?

In 2004 the College of Arts and Sciences was renamed the New College of Interdisciplinary Arts and Sciences (NCIAS). In late 2007 NCIAS was restructured, and instead of six departments and two programs it is now has three divisions: Humanities, Arts and Cultural Studies; Math and Natural Sciences; and Social and Behavioral Sciences. The Women's Studies faculty opted to move to the Division of Humanities, Arts and Cultural Studies because we believe it is a more conducive space to maintain a feminist scholarly environment where we can continue our scholarship based on innovative theories and methodologies. This new configuration will start in January 2008, therefore, it is too soon to know how it will work and what the implications for Women's Studies will be. However, there are concerns about the future viability of the academic departments, including Women's Studies, as in the new configuration there are no chairs of departments and programs. The majors (degrees offered) will probably remain untouched.

The way in which the college was restructured was surprising as few could foresee that chairs would be removed from their duties and that departments and programs would be absorbed by larger units. Saying that, a few faculty members in Women's Studies and Ethnicity, Race, and First Nation Studies were aware that their respective small size was a problem. At the end of the spring 2007 semester, a proposal to form a department between Women's Studies and Race, Ethnicity, and First Nation Studies was discussed among the faculty in these two programs. Women's Studies was not able to gain consensus on the proposal, therefore it did not move forward. This plan was circulated while the college was searching for a new dean. Given that it was not moved forward, it is not possible to know if this would have helped. From my perspective, at least it was an effort that organically originated with faculty and could have placed us in a stronger position.

The problem of Women's Studies size needs to be analyzed vis-à-vis the old organization of the college and in light with new reorganization.

The college from the outset was organizing interdisciplinarily (McGovern, 2007). The traditional departments of a College of Arts and Sciences such as English, History, or Sociology do not exist. The departments in NCIAS were organized around interdisciplinary areas such as American Studies (which became Language, Cultures and History), Social and Behavioral Studies, Interdisciplinary Arts and Performance; Integrative Studies, Applied Computing and Math, and Integrated Natural Sciences. And two programs: Women's Studies and Ethnicity, Race, and First Nations Studies. Women's studies has functioned as a department, even though it did not have the status, including having full-time faculty lines with their tenure home in Women's Studies. Ethnicity, Race, and First Nation Studies, from the outset relied on faculty from other departments even though it does offer a major. Moreover, in addition to the interdisciplinary degrees, the departments and the college did offer disciplinary degrees. Unlike most universities, the departments offered several majors, not only the one that was associated with the name of the department (e.g., history offering history degrees). Women's Studies' size made it susceptible to scrutiny, as institutional questions about why we should exist, when there wasn't a history department, for example, were constantly raised. Resources also became an issue, as some question why should the University support a program that has so few majors (but ignoring all the students we teach). In short, instead of being looked at as an asset to the University and an example of social and community embedded scholarship, the institution defined us as a problem. For example, a former dean repeatedly referred to Women's Studies as a "controversial" area of study.

In the introduction to the anthology *Women's Studies for the Future*, Elizabeth Lapovsky Kennedy and Agatha Beins (2005) propose to grapple with the implications of women's studies successful institutionalization. As I reflect on the experience of Women's Studies at ASU's West campus, I cannot but think that "we are not there yet." We have produced a good program with innovative curriculum, excellent faculty who are state-of-the-art teachers and scholars, and who have a strong commitment to social, gender, racial, and justice. In spite of our accomplishments, I don't think we have achieved the institutional stability that Kennedy and Beings allude to. We are also a program that throughout its history has consistently had to justify its existence. Although this is probably the history of many women's studies programs and departments, today, as demonstrated by the increase in graduate programs both at the Ph.D. and master's levels, and the number of students served institutionalization and stability is more often the case than not. In the case of Women's Studies at the West campus, while it is institutionalized, if suffers from its structural location.

Conclusion

bell hooks' wrote in the preface of the first edition of her influential book *Feminist Theory: From Margin to Center* that "to be in the margin is to be part of the whole but outside the main body" (xvi). As a faculty member in Women's Studies at the West Campus I have constantly felt that the program was marginalized. In reality, it was not necessarily marginalized as such, as it is inside the main body of the institution. I believe that Women's Studies position is akin to that of outsiders/within (Collins, 1998) living a borderland existence (Anzaldúa, 2007; Elenes, 1997), which is being inside the organization but not as full-fledged citizens. Women's Studies at the West Campus and its faculty have learned to navigate the institution and survive. The development of a vision that highlights the intersection of race, class, gender, and sexuality in national and transnational contexts has not been a contested terrain within the program. And until recently, we did not have to justify or explain the significance of focusing on scholarship that centralizes gender or race. That is, our problematic structural location enabled the possibility of enacting ideological vision, pretty much on our own terms.[8]

But being outsiders/within has serious limitations, as the structural location within the University and the college has put Women's Studies at a major disadvantage. Since I got to the campus, rumors have flown that the program will be closed or that we will be absorbed by a larger department. Oftentimes these rumors fell on deaf ears, but on too many occasions they created unneeded anxieties. As of this writing, we have been incorporated into a larger unit. Discussion of the formation of committees within this unit have ensued, and one more time the Women's Studies faculty are faced with the dilemma of whether or not to serve in time-consuming committees. It is starting to feel like the more things change; the more they stay the same.

Oppositional movements have their ups and downs; there are victories and there are failures. Successes and failures are the result of complex circumstances, oftentimes not of the making of those who are struggling for change. That is, individuals and groups in struggle can do everything in their power and ability to achieve their goals, and still fail. This is not meant as a pessimistic or defeatist statement but as recognition of the difficulties of the struggle against powerful forces and dominant ideologies. In the case of Women's Studies, we made possible quite a bit without much; we worked with what we had. This is not to say that we did not make mistakes, as we had our share of missteps. One in particular is that we let the institution define us as a problem and we reacted to the administrator's definition of Women's Studies that highlighted our deficiencies at the expense of our successes. As we were in constant struggle for the survival of the program, our voices got lost; we could have been more creative in front of the administration, and showed all the positive things we were doing and our level of productivity.

Nevertheless, women's studies in the United States has achieved a healthy level of institutionalization. Yet, caution is necessary and advocates of women's studies should not let their guard down. In an era when the academy is following a corporate model where "profits"[9] are privileged, institutions of higher education can shy away from their commitment to transformative education, including women's studies. But if universities are to continue their democratic and transformative educational mission, it is imperative that they look beyond mere numbers. It is important for colleges and universities make decisions based on values that promote access, excellence, maintain ties with different communities, and above all, maintain a commitment to social justice.

Notes

I want to offer a word of gratitude to Alice E. Ginsburg for her generous invitation to contribute to this anthology, and for her patience and support. Special thanks to Gloria Cuádraz and Luis Plasencia for their insightful comments in earlier versions of this chapter. Nevertheless, I am responsible for the analysis, representations, and possible slippages in this writing. I thank my colleagues old and new who have toiled and navigated with me the success and failures in Women's Studies. This chapter would have not been written without the support of my family, Manuel de Jesús Hernández-G, H-Tubtún, and Xchel.

1. Throughout this chapter I will use women's studies in both upper and lower case. Upper case is used when I specifically speak on the Women's Studies Program were I work, and lower case to refer in general to women's studies as a field of study.
2. During the early stages of development of ASU West colleges and departments could not be called as such. Colleges were divisions, and departments were programs. Therefore, the program heads were called coordinators who functioned as chairs, and divisions had directors who had the function of a dean. In the early 1990s, this changed and divisions became colleges with deans, and departments with chairs. Women's Studies literally fell through the cracks and remained a program, even though it has always functioned as

a department. A situation that has been very problematic since the late 1990s, making the program vulnerable to cuts and disenfranchisement.

3. Casanova offers an excellent analysis of the development of Ethnic Studies at the University of Wisconsin-Madison.

4. For an analysis and critique of the WOCC project in particular and curriculum transformation projects in general see Elenes (1995).

5. Importing literally means bringing in courses and requirements from one campus to another; the West campus has done this on several occasions when it was necessary to develop and implement curriculum quickly. One of these occasions was when the campus started to offer courses in 1980s, another one was when the campus added lower division and needed lower division courses in 2000. Women's Studies basically had the same curriculum as the Tempe campus until 1994. Once the faculty was on board at West, we were able to revise the curriculum according to our vision.

6. The term third world women is contested terrain, as it implies a hierarchy between the first and third world. I am using it here in the way that Chandra Mohanty (2003; 2005) and Chela Sandoval (1990; 200) respectively use it, to show the solidarity between women of color in the United States (and other so-called advanced nations) and the third world.

7. The low of 10 majors has been contested by Women's Studies because these number did not take into account double majors.

8. For many years, out tenure document gave more weight to chapters in books than journal articles because we believed that chapters in books have more dissemination than articles. This is not longer the case. Our recently approved tenure and promotion guidelines emphasize refereed journal articles, preferably in women's studies journals.

9. In institutions of higher education, profits come from external grants, patents, and partnership with corporations (and donors) that invest in the institution. As public funds shrink, universities seek alternative forms of funding via grants and corporate investment. But as any entity that depends on corporate monies, universities can start to worry about "bottom line issue" and shy away from its larger educational function. Students are also a fund of funds, even in public institutions where in-state student's tuition is subsidized.

References

Allen, P.G. (1986). *The sacred hoop: Recovering the feminine in American Indian traditions*. Boston, MA: Beacon Press.

Anzaldúa, G. (2007). *Borderlands/La Frontera: The new Mestiza*. 3rd edition. San Francisco, CA: Aunt Lute Books.

Asian Women United of California (Eds.) (1989). *Making waves: An anthology of writings by and about Asian American women*. Boston, MA: Beacon Press.

Baca Zinn, M., L. Weber Cannon, E. Higginbotham, and B. Thorton Dill. (1986). "The costs of exclusionary practices in women's studies." *Signs: Journal of Women in Cultures and Societies*, 11(2), 290–303.

Brown, M. and M. Chávez-García. (2005). "Women's studies and Chicana studies: Learning from the past, looking to the future." In E.L. Kennedy and

A. Beins (Eds.), *Women's studies for the future: Foundations, interrogations, politics*. New Brunswick, NJ: Rutgers University Press, 143–155.

Casanova, S. (2001). The Ethnic Studies movement: The case of the University of Wisconsin-Madison. Unpublished doctoral dissertation. University of Wisconsin-Madison.

Collins, P.H. (1998). *Fighting words*. Minneapolis: University of Minnesota Press.

———. (2000). *Black feminist thought*. 2nd edition. New York: Routledge.

Davis, A. (1981). *Women, race, and class*. New York: Vintage Books.

Elenes, C.A. (1995). "New direction for feminist curriculum transformation projects." *Feminist Teacher*, 9(2): 70–75.

———. (1997). "Reclaiming the borderlands: Chicana/o identity, difference, and critical pedagogy." *Educational Theory*, 47(3), 359–375.

Guy-Sheftall, B. and E.M. Hammonds. (2005). "Whither black women's studies: An interview." In E.L. Kennedy and A. Beins (Eds.), *Women's studies for the future: Foundations, interrogations, politics*. New Brunswick, NJ: Rutgers University Press, 61–71.

hooks, b. (1984/2000). *Feminist theory: From the margin to center* 2nd edition. Cambridge, MA: South End Press.

Kennedy, E.L. and A. Beins. (2005). "Introduction." In E.L. Kennedy and A. Beins (Eds.), *Women's studies for the future: Foundations, interrogations, politics*. New Brunswick, NJ: Rutgers University Press, 1–28.

Ladson-Billings, G. (2000). Racialized discourses and ethnic epistemologies. In N.K. Denzin and Y.S. Lincoln (Eds.), *Handbook of Qualitative Research*. 2nd edition. Thousand Oaks, CA: Sage, 257–277.

McGovern, T. (2007). Academic pluralism, faculty development, and interdisciplinary curricula: Continuing the narrative for the New College of Interdisciplinary Arts and Sciences. Unpublished paper.

Mohanty, C.T. (2003). *Feminism without borders*. Durham, NC: Duke University Press.

———. (2005). "'Under western eyes' revisited: Feminist solidarity through anticapitalist struggles." In E.L. Kennedy and A. Beins (Eds.), *Women's studies for the future: Foundations, interrogations, politics*. New Brunswick, NJ: Rutgers University Press, 72–96.

Moraga, C. and G. Anzaldúa. (Eds.) (1983). *This bridge called my back: Radical writings by women of color*. 2nd edition. New York: Kitchen Table, Women of Color Press.

Sandoval, C. (1990). "Feminism and racism: A report on the 1981 National Women's Studies Association Conference." In G.A. Anzaldúa (Ed.), *Making face, making soul: Haciendo caras*. San Francisco, CA: Aunt Lute Books, 55–71.

———. (2000). *Methodology of the oppressed*. Minneapolis: University of Minnesota Press.

10

Women's Studies, Health, and Science: Evolution from Separation to the Beginnings of Integration

Sue V. Rosser

Much has changed during the almost 35 years that I've been involved with women's studies. The demographics and views of the students, the growth from a few undergraduate courses scattered in a handful of institutions to hundreds of programs, including some with Ph.D. degrees, the expansion of research and plethora of texts, and increased emphasis upon diversity and international perspectives typify some of the changes. Despite the significant growth and changes, writing this essay made me remember that the two major issues that initially drew me decades ago to women's studies still remain: (1) women having knowledge and control over their own bodies and (2) building two-way streets between women's studies and science to get more women and feminist perspectives into science and technology and more science and technology into women's studies. Although the ways we frame them and some of the vocabulary we use to discuss them may differ today, the fundamental issues remain.

My Initial Involvement in Women's Studies

Experiences over reproductive issues, body control, and career served as personal motivations for my initial involvement in women's studies. In the mid-1970s, after two years of learning a new area of biology, my postdoctoral research finally appeared to be on track. When I became pregnant with my second child, the professor supervising my research suggested that I get an abortion, since it was "the wrong time in the research and we needed to obtain more data now to have the grant renewed." I did not have the abortion; I rationalized that taking minimal time off for childbirth meant that everything would be fine in the lab and with my scientific career. One day when I was out at noon to breastfeed the baby, a call came that focused on my area of research. Although many of the other postdocs and graduate students (all males) used their lunch hour to play squash, I later learned that the professor had made comments to others in the lab about my being off nursing again. Those comments coupled with related incidents made me decide to accept the offer of the new women's studies program to teach a course on the biology of women.

Women's studies gave me new space, perception, knowledge, and connections to understand my bodily identity and the reactions it invoked. In the mid-1970s, only women were involved in this first year of the program; they welcomed me because I was a woman and even encouraged me to bring the baby to some meetings. From the stories of the more senior women in the academy, I learned much of the discrimination I had experienced was not unique, but resulted from being a woman in a patriarchal university that only the previous year, under court order, had dropped its official quotas on women medical students and nepotism rules against women faculty. From teaching biology of women, I began to read the evolving critiques revealing bias in research developed by men, using only males as subjects, and with theories and conclusions extrapolated inappropriately to the entire population of both men and women. While developing a new course in women's health, I recognized that despite having received by Ph.D. in zoology and having given birth to two children, I knew almost nothing about my own body and its functioning. When I sought materials to fill this knowledge gap, I learned that very little research had focused on women and their bodies in health and disease. Recognizing this dearth and that any research that existed had been undertaken by men from their perspective, I began to work to create materials that reflected women's experience of their biology and bodily identity. Suddenly I felt a connection with teaching this research and with my search for materials from medical textbooks through scientific journals to fiction to convey the information to students. Trained in the humanities, social sciences, and fine arts, my colleagues in women's studies helped me to uncover the interdisciplinary resources needed to understand the components of women's health.

My Mentor

Ruth Bleier, a founder of the women's studies program at University of Wisconsin-Madison, was a pioneer who helped develop and co-taught "Biology and Psychology of Women," one of the first four core courses in the program. Imbued with a passionate vision of the struggle for peace, for feminism, and for truth, and justice, Ruth continued to be a guiding light for women's studies at the university, where she served as chair of women's studies from 1982–1986, and for feminists throughout the world, where her scholarship on feminism and science became well known.

Ruth Bleier was the one individual who, more than any other person, opened up for me the possibility of combining feminism and science. In my experience as a biologist I found that many scientists were feminists in their politics or in their relationships outside of the laboratory. Although they fought for feminist causes in other arenas, they left feminism at the laboratory door, divorced from their scientific hypothesizing, data gathering, and theorizing. Some scientists who were feminists, then questioned whether or not gender entered into good science; this attitude still remains.

Of course most feminists were not and are not scientists. Certainly in the 1970s and even today, a quick perusal of the women's studies journals and disciplinary affiliations of most feminists in academia yields an overwhelming majority of scholars in the humanities and social sciences. Few feminists outside of academia are scientists. In fact, many feminists openly reject science, which they continue to view as masculine, dominating, patriarchal approach to the world and women.

Ruth Bleier represented one of a small group of individuals interested in both science and feminism, and the interaction between the two. She spent hours explaining the connections between feminism and science to her colleagues and friends who were not scientists. She used her feminist analysis to critique existing theories of science, to point out racist and sexist flaws in experimental design and interpretation, and to begin to sketch the parameters for a feminist science. Perhaps most importantly, she brought the feminist critique to bear on her own research and that of her colleagues in neuroanatomy.

As an administrator, Ruth Bleier served as a mentor and a role model in a way that I tried to emulate when I assumed administrative positions. She continued to be a practicing scientist while working on and writing about feminism and science during the four years that she chaired the women's studies program. Ruth continued active neuroanatomy research, well supported by federal grants, while also writing on feminism and science. Her best known single-authored book *Gender and Science* and a monograph on the cat brain were published simultaneously in 1984. Over the years, I've drawn inspiration from Ruth's model at times when my duties as a women's studies director for more than 20 years at three different institutions and for the past nine years as dean have made it a struggle to keep my research going.

Ruth was a feminist scientist who could most skillfully use the methods and theories of feminism to critique science and the tools of science to analyze flaws in the methodology and theoretical constructs in feminism. Those skills have proved invaluable as ways to use women's studies to radicalize science and health and to use science and health to radicalize women's studies as both have evolved.

Women's Studies as a Radicalizing Force

A quick look at interactions between women's studies and women's health during the last four decades provides insights into their co-evolution. At critical times a women's health issue such as abortion, contraception, or hormone replacement therapy galvanizes research and activism in women's studies. Similarly, new paradigms, theories, or emphases in women's studies feed research and policy in women's health. A force motivating the wave of feminism begun in the 1960s was women's health concerns. Women sought the right to control their own bodies through access to safe birth control, abortion, and information about their physiology and anatomy, to define their own experiences as a valid aspect of their health needs, and to question the androcentric bias found in the hierarchy of the male-dominated health care system and its approach to research and practice.

Activist individuals, encouraged partially by the women's movement in the late 1960s, brought legal challenges, including class action suits, against medical schools that had quotas on the numbers of women, as well as racial and ethnic minority men, admitted. Their legal challenge of the quota system stood as one of the first of many critiques women presented to the medical profession. Women activists can claim responsibility for initiating a demographic shift within the medical profession that increased the percentage of women medical school graduates from about 7 percent in the mid-1960s to around 50 percent today.

Not until a substantial number of women had entered the professions of biology and medicine could biases from androcentrism be exposed. Once the possibility of androcentric bias was discovered, the potential for distortion on a variety of levels of research and theory was recognized: the choice and definition of problems to be studied, the exclusion of females as experimental subjects, bias in the methodology used to collect and interpret data, and bias in theories and conclusions drawn from the data. Since the practice of modern medicine uses a biomedical approach based in positivist research in biology and chemistry and depends heavily on clinical research, any flaws and ethical problems in this research are likely to result in poorer health care and inequity in the medical treatment of disadvantaged groups (Rosser, 1994).

Not surprisingly, controversies in breast cancer research demonstrate many of the problems that women's health, in general, has suffered at the hands of a male-dominated, hierarchical health system that is based on a biomedical model of medicine. The biomedical model focuses on anatomy and physiology and causes of disease at the cellular, hormonal, and genetic levels rather than behavioral, social, and environmental contributions to disease. Since breast cancer is impacted by behavioral and environmental factors and is not a major health problem for men, it has received low priority, funding, and attention.

Women scientists, consumers, physicians, and politicians brought these revelations and other examples of bias and gaps in research and practice to the attention of the health community. In 1991, Bernadine Healy, M.D., established the Office of Research on Women's Health and announced plans for the Women's Health Initiative. The Women's Health Initiative, designed to collect baseline data and look at interventions to prevent cardiovascular disease, breast and colorectal cancer, and osteoporosis, seeks to fill the gaps in research and practice.

Expanding the Definition of Women's Health

The interdisciplinary nature of most issues in women's health and disease formed the raison d'être for health care professionals and researchers from diverse specialties, professions, and disciplines to interact. They recognized that the traditional territory of obstetrics/gynecology failed to include much of women's health beyond the defined limits of reproductive issues and they expanded women's health to include differences in frequency, symptoms, and effects of diseases found in both sexes. The expanded definition also requires linkages with researchers in liberal arts to understand women's health. Behavioral and social sciences, particularly psychology, sociology, anthropology, and women's studies, provided basic research theories for domestic violence, eating disorders, and sexual and/or physical abuse, now recognized as central to women's health.

Impact upon Curriculum and Research

Part of the impetus initiating the women's health movement was women's desire to have access to information about their physiology and anatomy as part of the need to control their own bodies. Women's health courses sprang up in women's studies, nursing, sociology, and in a variety of places in the curriculum, beginning in the early 1970s in an attempt to quench that desire. In a 1986 volume, I described my reasons for developing and teaching such courses to fill the void in the women's studies curriculum:

> The goal of Women and their Bodies in Health and Disease is to help women regain control over their bodies through learning the language and understanding the theories of science and health...Women and Their Bodies in Health and Disease is an introductory course without any prerequisites. For most students it may be their first and only college-level science or health course; frequently women students take this course to fulfill their natural science requirements because they are too afraid of a "regular" science course. For many students, it may also be their first women's studies course. In brief, this course is likely to attract a group of students who feel very unsure about their science ability, very unwilling to question authority, and rather fearful of talking about these issues. (Rosser, 1986, 84)

Empowering women through providing them with scientific vocabulary and some knowledge of their bodies stood out as strong motivations for my offering this course.

At that time, although topics such as women and exercise, nutrition and diet, stress, and mental health, diseases of women, and aging and menopause were included in the course, I focused over half of the course on sexuality and reproductive issues. Over the years, while I continue to include most of the same basic topics, I've increased the time on wellness, menopause, care-giving for elders, and challenges of dealing with the health care system. I wonder how much this shift expresses my own interest in these topics as I have encountered menopause, cared for my aged mother until her death, and dealt with the shifting face of U.S. health care. Indeed, upon occasions during the course, a student has expressed to me that she now understands better what her mother is going through in ways she had not previously because of her own preoccupation with coming out or problems with contraception.

Diversities among Women

Even in the late 1970s and early 1980s, diversities among women helped to shape discussions of topics in women's health. An example of discussion of the relatively "new" technologies in my 1986 book, *Teaching Science and Health from a Feminist Perspective* illustrates one of the ways I included diversities among women:

> After a review of the biological events of a normal pregnancy, I introduce the various technologies frequently used during pregnancy. I explain the biological principle surrounding amniocentesis...After reading and discussing the biology and technicalities of the procedure, I move to a consideration of the technology's benefits...Then I explore some of its risks....When the students consider amniocentesis from the perspectives of class and religion, they begin to realize that since it may be expensive, at this point, it is restricted mostly to middle-class women whose religious background has limited sanctions against abortion. After some reading (Saxton, 1984) and a guest lecture by a physically challenged woman, students broaden their perspective to recognize that the issue of amniocentesis raises questions about what it means to be labeled *abnormal* or defective in our society and about who decides who is "normal," therefore worthy of life, and who is "abnormal," and should be aborted. Women of color point out that other tests of normality (IQ) have often been used to screen out non-whites and represent people of color as inferior (Chase, 1980). (Rosser, 1986, 41–42)

Two major features now distinguish the ways I discuss reproductive technologies compared to the late 1970s and early 1980s. First, many more technologies and much more complexity in the ways they can be used exist in the twenty-first century. Second, the significance of international impacts of globalization and their intersection with ethnicity, class, sexualities, and religions, broaden the diversities further.

I now focus the discussion around how reproductive technologies have different impact upon people of different classes and in different countries. The capabilities opened by in vitro fertilization have led to an industry in which poor women are paid to gestate the developing fetus, who may represent the genetic offspring of a rich man and a woman who does not wish to and/or can't carry a pregnancy to term—the so-called Baby M phenomenon. On a worldwide scale, particularly in India and China, amniocentesis is used for sex selection to abort more female than male children because traditionally in those cultures, sons are valued more, partially because of their economic contributions to the family income and to the care of their elderly parents. A person born in an overdeveloped country consumes about 30 times as much as a person in an underdeveloped country. Despite the fact that children born in industrialized, first world countries, especially middle-class ones, are encouraged to have children, women in developing countries may be sterilized against their will. Infertility becomes the issue in first world countries, while fertility signals the problem in developing countries.

Women's Studies and Science

A similar co-evolution of women's studies and science has occurred, with considerably less far-reaching impact on either field. Although feminist critiques of science and efforts to remove barriers to women in science, technology, engineering, and mathematics (STEM) also began in the 1960s, in fields such as computer science, engineering, and the some of the physical sciences, women are far from parity, constituting less than a quarter of scientists. The bulk of the science curriculum and research agenda remains untouched by women's studies. Simultaneously, very few women's studies programs have a major or significant concentration in feminist science studies or women in science, outside of women's health. Many programs have no scientists as core or tenure-track faculty in women's studies. This dearth of scientists resulted partially from the very small numbers of tenure-track faculty women in senior and leadership positions in science, technology, engineering, and mathematics. Strong cultural traditions of masculinity and objectivity in science threatened to keep women's studies separate from the theories of cultural and social construction of knowledge production acceptable in the humanities and social sciences.

In 1988, I expressed the loneliness I felt at being one of so few people focused on women and science: "For years I have always felt an outsider at national professional meetings in either science or women's studies" (Rosser, 1988, 105). Now, most campuses boast women in science and engineering (WISE) programs for students, each year numerous conferences, journals, and anthologies focus on women and science, and the National Science Foundation and other federal agencies award multimillion dollar grants to facilitate institutional transformation to advance and retain women in science and engineering.

In many ways, the scholarship on women and science mirrors the categories of scholarship in women's studies as a whole and the emerging development of the field.

History of Women in Science

Recovery of lost texts and figures characterize some of the earliest scholarship in the late 1960s and early 1970s as women's studies emerged as the academic arm of the women's movement with the establishment of the first programs in 1969–1970 (Hedges, 1997). The search for where and why women were missing from all fields was a necessary first step in beginning to understand how their absence led to flaws, distortions, and biases in each discipline. History of women in science and their impact upon the different disciplines and subfields continues to be an active research area today.

Current Status of Women within the Professions

Recognition that basic data on the numbers of women relative to men receiving degrees in science, mathematics, and engineering, and their employment status, rank, salary, and professional progress and attainments were crucial to women and science came early. After a successful lobby of Congress, the Science and Engineering Equal Opportunities Act of 1980 was passed. The National Science Foundation was required to collect data each year on the status of women and other underrepresented groups; in the 1990s the data collection included persons with disabilities in science.

Building on these foundational data, current scholars provide statistical documentation and analyses of more subtle factors and obstacles that now deter women. The dearth of women faculty can be traced only in part to relatively small numbers of women graduating with undergraduate and graduate degrees in many areas of the sciences, engineering, and mathematics. Issues surrounding chilly climate, competition between the biological clock and the tenure clock in rapidly evolving disciplines in which presence in the laboratory or field is needed to undertake research, as well as isolation, lack of camaraderie and mentoring, and difficulty in gaining respect from peers cause attrition at all levels of the science pipeline (Rosser, 2004).

Inclusions and Exclusions: Gender Differences and Diversity among Women

Just as women's studies scholars revealed that the assumption that male experience coincided with human experience constituted a form of androcentric bias that rendered women invisible and distorted many research results, these same scholars mistakenly assumed that the experience of all women was the same. Women of color, working-class women, and lesbians pointed out that their experiences as women and as scientists did not fit the depictions that emanated from a white, middle-class, heterosexual perspective. This revelation led to the recognition that gender did not represent a homogeneous category of analysis and that gender needed to be studied in relationship to other oppressions of race, class, nationalism, and sexual orientation. Age or developmental stage becomes another aspect of diversity that can modify the experience of even the same woman throughout her life course.

During the past 15 years, we have begun to recognize the influence of globalization and the significance of understanding international perspectives and movements. Much in the same way that early on, in its eagerness to discover the influence of gender, women's studies suffered from the failure to recognize diversity among women, scholars now acknowledge the constraints of not understanding the experiences of women in different countries and cultural contexts. Although nearly 30 percent of the actively employed science and engineering doctorate holders in the United States are foreign born, as are many postdocs (NSB, 2004), very little research has focused on immigrant women scientists. One study (Xie and Shauman, 2003) found that immigrant women are only 32 percent as likely as immigrant men scientists and engineers to be promoted, partly because the women tend to immigrate for their husband's career.

Revealing Male Subtexts and Building Alternative Models

As women's studies entered a stage that focused on the analysis of gender as a social category, critics began to question the ways in which gender determines the structure of social organizations, systems of cultural production, and the roles and definitions of masculinity and femininity. Scholars explored how the scientific hierarchy, including the language and metaphors of scientific theories and descriptions used, both reflected and reinforced gender roles. They uncovered the historical roots of modern science in a mechanistic model in which objectivity became synonymous with masculinity and that encouraged the domination of male scientists over women, nature, and organic models of the world.

Theory into Practice

Many scientists (Gross and Levitt, 1994; Koertge, 1994) rejected the postmodernism espoused by their colleagues in the humanities and social sciences. Many women scientists and engineers, while appreciating the issues raised about objectivity, questioned the translation of "high theory" into practice of science and the relevance for such theories in their own lives as scientists (Koertge, 1994), where they still encounter substantial discrimination. The science wars that developed from postmodern theories and increasing globalization drew attention to the necessity for the refusion of theory and practice. For many women teaching and practicing science, this dichotomy between theory and practice appeared to be a false separation. Grounded in laboratory practice, the fusion of theory and practice in science classrooms laboratories has a long tradition.

Feminist Science Studies in the Daily Lives of Women Scientists

Further evidence of the fusion of theory with practice comes from a current focus of feminist science studies on the personal experiences and daily lives of women scientists. These studies also reflect interdisciplinary approaches in their use of postcolonial theories, oral histories, and ethnographies as theoretical and methodological approaches to science studies. Would women's differing interests, life experiences, and perspectives lead them to ask new questions, take different approaches, and find alternative interpretations leading to new theories and conclusions?

The genre of biography and the methodologies of qualitative research have witnessed increased enthusiasm with both the public and scholarly communities. A continuing theme of women in science, women's studies scholarship, and feminism has focused on grounding the scholarship in women's personal experiences.

Integrating Feminism and Women's Studies into the Science Curriculum

Including curricular content on women in science and feminist science studies in the science curriculum has proved challenging. Although in the 1970s and 1980s feminist science studies did not exist and the "culture wars" over theory had not yet occurred, many science departments resisted the notion that courses focused on gender belonged in the science curriculum.

In 1986, I looked back upon the issues and dearth of models for teaching a women in science course in the late 1970s and early 1980s:

> Women in Science: History, Careers, and Forces for Change is the only course at the institution where I teach that attempts to deal with pragmatic problems of careers, the history of women scientists, or the changes that might occur in traditional scientific theory if considered from a feminist perspective. Although some large institutions have history of science programs, containing courses or sections of courses devoted to women, rarely are considerations of career strategies and the ultimate changes in theories included in these courses. I suspect that most institutions would have only one course dealing with these matters, if they have any at all. This means that a very broad area of content must be considered in a single semester. The history, career strategies, and potential of women as forces for change in all of the physical and biological science, mathematics, computer science, and medicine include a huge spectrum of information. (Rosser, 1986, 65–66)

Because of the very few courses focused on women and science, no textbooks existed. Piecing together appropriate texts and materials from a variety of sources constituted a struggle:

> *Texts:* Increasing numbers of books, which might serve as texts for the course appear each year. Originally, when I began teaching the course, only Haber's *Women Pioneers of Science* (1979), Osen's *Women in Mathematics* (1974), Mozan's *Women in Science—1913* (1974), and Sayre's *Rosalind Franklin and DNA* (1975), were available. I supplemented these with journal articles. Now, several excellent works are also available: Rossiter's *Women Scientists in America* (1982), Gornick's *Women in Science* (1983), Goodfield's *An Imagined World* (1981), and Keller's *A Feeling for the Organism* (1983). Last year, I successfully used the Haber book as a general text, with the Gornick and Keller books providing individual portraits. In addition, I used journal articles or *Science and Liberation* (Arditti, Brennan, and Cavrak, 1980) for their feminist critiques of traditional science. Bleier's *Science and Gender* (1984) or *Alice Through the Microscope* (Brighton Women and Science Group 1980) may also serve this purpose. (Rosser, 1986, 69)

Now at the technologically focused institution where I serve as dean, we have an entire minor focused on women, science, and technology

(Rosser, Fox, and Colatrella, 2002). Teaching the course in the twenty-first century that most closely parallels the Women in Science course I taught in the early 1980s, makes me realize the considerable progress of the field. First, the title of the course—Feminist Analyses of Science and Technology—includes the word feminist. In the course objectives, we underlined the importance of feminist theories:

> In this course you will become familiar with a range of feminist theories as they apply to science and technology. You will find that not all feminists think alike—indeed, the approaches and goals of various feminist theories differ greatly. Learning these distinctions will enable you to become a sophisticated analyst of issues concerning science and technology. (HTS 4011 Syllabus)

Many possibilities for texts now exist. Each subfield of the course has numerous recent, excellent volumes that can be assigned or from which chapters may be used. In addition to the large numbers of texts focused on the history of women in science, books such as *Women in Science* (Xie and Schauman, 2003), *The Science Glass Ceiling* (Rosser, 2004), and *Why Aren't More Women in Science* (Ceci and Williams, 2007) provide quantitative data to analyze the structural issues and career barriers for women scientists. *Cyberfeminism* (Hawthorne and Klein, 1999), *Biopiracy* (Shiva, 1996), and *Evolution, Gender, and Rape* (Ed. Travis, 2003) permit more in-depth studies of particular areas. *Is Science Multicultural?* (Harding, 1998) and *Has Feminism Changed Science?* (Schiebinger, 1999) serve as single-authored explorations of the impact of feminism on science, while *Gender and Technology* (Ed. Lerman, Oldenziel, and Mohun, 2003), *Women, Gender, and Technology* (Fox, Johnson, and Rosser, 2006), *WildScience: Reading Feminism, Medicine and the Media* (Marchessault and Sawchuk, 2000), and *Feminism in Twentieth Century Science, Technology, and Medicine* (Creager, Lunbeck, and Schiebinger, 2001) constitute just a few of the many anthologies of articles appropriate for textbooks for courses conjoining feminism and science.

Women's Studies and Science Join to Change the Academy

After almost four decades of attempts to build two-way streets, women's studies and science have begun to merge in some national initiatives. Science, technology, and health have had some impact on women's studies scholarship, as well as courses. An indicator of the small impact of science on women's studies emerged from analyses I did from 1979 to 2001 of the number and percentage of papers and entire sessions on science, technology, computer science, or health presented at NWSA annual conferences during those years. The percentage of papers ranged from 5.5 to 16.77 percent in 1979–1986 and from 7.0 to 19.60 percent in 1987–2000; the percentage of sessions ranged from 4.1 to 14.07 percent in 1979–1986 and from 4.5 to 16.00 percent in 1987–2001.

Recent initiatives at the National Science Foundation (NSF) hold the potential to radically transform institutions. I was personally pleased that a conference that NSF asked me to host to bring together scientists from the community with program officers from NSF lead to the recommendation for institutional structural changes in universities to facilitate advancement of women faculty to senior positions. NSF ADVANCE program has primarily limited its support to increasing the participation of women in science and transformation of the structures of institutions to make them more accessible and friendly to women scientists; it has not encouraged reconceptualization of research to put women and gender in central focus.

In contrast, the National Institutes of Health (NIH) has begun to emphasize reconceptualization of research to include women and gender in its focus and analysis of results, including gender-based medicine. NIH put into effect new guidelines for Phase III clinical trials covering both the inclusion of women and sex-based analysis for reviewers and scientific review administrators. In April 2001, the Institute of Medicine published *Exploring the Biological Contributions to Human Health: Does Sex Matter?* (Wizemann and Pardue, 2001). The agenda validated the study of basic biologic and molecular bases for sex and gender differences in disease, including "sex-based biology as an integral part" of research conducted by the Institutes.

The differences in the approaches of these two agencies suggest that NIH and NSF might learn from each other. I was pleased to participate in a plenary session at a November 2007 NIH Conference on Mentoring Women in Biomedical Careers. On the basis of NSF's experience, NIH is considering institutional transformation to insure the impact of advancing women to decision-making positions impacting research. Simultaneously, NSF must begin to reconceptualize research in science to focus on women and gender. This reconceptualization constitutes a more difficult proposition in basic research than it does in the applications of clinical medicine where gender is obvious. Both NSF and NIH have acknowledged the importance of work in women's studies and have attempted to include women's studies scholars into their initiatives.

These same issues, women's knowledge and control of their bodies and women, feminism, and science, that attracted me to women's studies 35 years ago have become even more salient today. Now framed as gender-based medicine and institutional transformation, the new rhetoric brings the major health and science funding agencies of NIH and NSF together with fundamental women's studies initiatives to place women and gender at the center of research agendas and transformation of educational institutions.

References

Bleier, R. (1984). *Gender and science*. Elmsford, NY: Pergamon Press.

Ceci, S.J. and W.M. Williams. (2007). *Why aren't more women in science*. Washington, DC: American Psychological Association.

Creager, A.N., E.A. Lunbeck, and L. Schiebinger. (Eds.) (2001). *Feminism in twentieth century science, technology, and medicine*. Chicago: University of Chicago Press.

Fox, M.F., D. Johnson, and S.V. Rosser. (Eds.) (2006). *Women, gender, and technology*. Urbana and Chicago: University of Illinois Press.

Gross, P. and N. Levitt. (1994). *Higher superstition: The academic left and its quarrels with science*. Baltimore, MD: Johns Hopkins University Press.

Harding, S. (1998). *Is science multicultural?* Bloominton: Indiana University Press.

Hawthorne, S. and R.D. Klein. (1999). *Cyberfeminism*. Melbourne, Australia: Spinifex.

Hedges, E. (1997). "Looking back, Moving forward." Editorial in *Women's Studies Quarterly* 25(1–2): 6–13.

HTS 4011 Syllabus. Taken from 2002 syllabus of class taught by I. Kennelly and S.V. Rosser.

Lerman, N., R. Oldenziel, and A. Mohun. (Eds.) (2003). *Gender and technology*. Baltimore, MD: Johns Hopkins University Press.

Koertge, N. (1994). "Feminist epistemology: Stalking an undead horse." *Annals of the New York Academy of Sciences* 775(1): 413–419.

Marchessault, J. and K. Sawchuk. (Eds.) (2000). *WildScience: Reading feminism, medicine and the media*. New York: Routledge.

National Science Board (NSB). (2004). *Science and engineering indicators 2004*. Volume 1. Arlington, VA: National Science Foundation (NSB 04–1).

Rosser, S.V. (1986). *Teaching science and health from a feminist perspective: A practical guide*. New York: Pergamon Press.

———. (Ed.) (1988). *Feminism in the science and health care professions: Overcoming resistance*. New York: Pergamon Press.

Rosser, S.V. (1994). *Women's health: Missing from U.S. medicine.* Bloomington, IN: Indiana University Press.

———. (2004). *The science glass ceiling.* New York: Routledge.

Rosser, S.V., M.F. Fox, and C. Colatrella. (Fall/Winter 2002). "Developing women's studies in a technological institution." In *Women's Studies Quarterly: Women's Studies Then and Now,* 30(3–4): 109–125.

Schiebinger, L. (1999). *Has feminism changed science?* Cambridge, MA: Harvard University Press.

Shiva, V. (1996). *Biopiracy.* Boston, MA: South End Press.

Travis, C. (Ed.) (2003). *Evolution, gender, and rape.* Cambridge, MA: MIT Press.

Wizemann, T.M. and M. Pardue. (Eds.) (2001). *Exploring the biological contributions to human health: Does sex matter?* Institute of Medicine. Washington, DC: National Academies Press.

Xie, Y. and K. Shauman. (2003). *Women in science.* Boston, MA: Harvard University Press.

11

Reflections on a Feminist Career

Nancy A. Naples

Women's Studies has been my formal academic home since 1992. After 15 years, I continue to find Women's Studies one of the most exciting, challenging, and important academic sites despite the development of new interdisciplinary units such as human rights, justice studies, science studies, and cultural studies. Women's Studies remains, for me at least, the most intellectually stimulating and politically significant academic location for bridging theory and political practice, and encouraging consciousness raising among a new generation of activists and scholars. As an interdisciplinary and praxis-oriented site committed to intersectional analysis and political practice, Women's Studies has allowed me to continue exploring, teaching, and writing about themes that shaped my early activism.

Linking the Personal, Political, and Academic

My development as a feminist and my location in the interdisciplinary academic field of women's studies is very much a story of praxis, a theme that pervades both my academic work, my teaching, and my personal politics. During the decade of the 1970s I worked with a diversity of women organizers throughout New York City, in part, through my role as director of community and family services for the YWCA of New York City and, more significantly, as a political activist concerned with protecting the rights and well-being of low-income adolescent parents with whom I worked. My decision to return to graduate school was prompted by my overload as a political activist and my frustration as a social worker—in the jargon of the social work field, I was suffering from "burn out." My dissertation research on women community activists in New York City and Philadelphia, many of whom had fought on behalf of their communities for decades, was a response to this personal dilemma. How did they keep going?

Many of the activist women with whom I worked in the 1970s labored with and through the community action programs funded through the antipoverty agencies of the Economic Opportunity Act of 1964. Although a cohort of these community workers were from middle-class backgrounds, a disproportionate number were from low-income and working-class backgrounds. There was high turnover among the more middle-class activists and social workers, while those who came to the community struggles as community residents and who self-identified with the racial-ethnic community represented were more likely to remain active over the decade. As I pondered the oft-asked question of how to keep people involved in community actions for social justice, I recognized the need to understand the complex ways that people became politically conscious and how they engaged in varying political actions before I could continue to be effective as a community organizer. I also became fascinated at the contradictory role that the state played in shaping the terms of the debates within the community-based struggles as well as providing sites for organizing against the state itself. But the guiding motivation for my research was a more deeply felt quest to find my own place in the numerous political movements for economic and social justice.

As of result of my research with African American women and Latina activists from low-income communities in New York City and Philadelphia (Naples, 1998b), I learned the significance of what Fran Peavey (1986) calls "heart politics." Given the structural impediments to social change and equity, one must reach into her heart and ground political efforts in personal goals and private desires. This perspective broadens the feminist insight that the "personal is political" by situating political efforts within personal visions of "the just society" in which we would like to live.

My research on women and poverty stems from my identification with those who are struggling. I have always understood their lives in terms of resistance. Working with low-income adolescent mothers in New York City, it was apparent to me how their lives were constrained by sexism and racism as well as poverty. Yet they were survivors. Listening to their stories, I was often amazed at their abilities to challenge, work the system, resist racist and classist definitions. But my research on the dynamics of gender, race-ethnicity, and poverty did not grow directly out of my own life experience. Although I identified so much with the challenges they faced as women, I am not a woman of color, I was not born in poverty. As a working-class woman, I found myself caught between struggles. The Women's Movement did not speak to a lot of my experiences, at least within the framework I had at the time. The Left was male-dominated and, therefore, I felt uncomfortable and often marginalized in that context. My journey as a political activist and feminist was fundamentally about finding a centered place where I could feel "at home." I found that home in women's studies, although, as in my family home, it was not always an easy place to be for a variety of reasons, some of which I will detail in this essay.

Feminism as an Identity

It wasn't until I entered graduate school that I was explicitly drawn into "feminism." Although I was active in community-based and New York City–wide struggles for equity for almost a decade, I could not relate to the liberal, white, middle-class women I met over the years who, I believed, represented the Women's Movement in New York City. I felt they were ineffectual in their political organizing and failed to speak to the needs of low-income women of color who formed most of my political network at the time. As a consequence of my own working-class background, I was particularly sensitive to the dynamics of classism.

Unlike the journey of many of my middle-class academic colleagues in Women's Studies, my feminism grew from identification with low-income women. I learned from their creative strategies of resistance that women can be both nurturing and strong. I watched how they spoke from their heart with great wisdom and courage and often was amazed by their patience and talents. I was particularly drawn to the African American women and Latinas who played a central role in the community-based struggles in New York City during the 1970s. I learned to be a feminist from their words and deeds. I owe a great deal of my feminist consciousness to them. And, not surprisingly, many of these women would not call themselves feminists. My notion of feminism goes beyond the limits of self-definition, it is something that oozes out of the most unlikely corners and is nourished by many who like "activist mothers" are "doing just what needs to be done" (Naples, 1998b) to secure the well being of those they love.

My journey toward feminism and into Women's Studies was furthered by the powerful presence of feminist scholars at the Graduate Center of the City University of New York (CUNY).[1] The interdisciplinary nature of feminist scholarship at the Graduate Center provided a fertile ground for many of us who were fortunate to study there in the 1980s. The feminist graduate and post-graduate student community offered an even richer basis for personal and intellectual empowerment.[2] My mentors and colleagues introduced me to academic feminism, which, in turn, provided me with the tools to revisit my previous work and reintegrate the lessons I learned from women community organizers from poor neighborhoods in New York City into an intersectional vision of feminist activism.

Feminist Analyses, Women's Issues, and Academic Products

Throughout the twentieth century and beyond, feminists scholars and activists have convincingly argued that feminism encompassed more than a concern with "women's issues." They demonstrated that capitalism, militarization, colonialism, poverty, environmental degradation, among other oppressions, must be understand through a gendered lens (see, e.g., Enloe, 2000). Feminists pointed out that women's social location placed them into particularly vulnerable positions with regard to the problems that result from large-scale economic, political, and social processes such as global economic restructuring (see, e.g., Beneria and Feldman, 1992). They simultaneously argued that those issues associated with women, such as childcare, reproductive rights, and food security, have profound effects on all members of households and communities regardless of gender (see, e.g., Stout, 1990).

My first book (Naples, 1998a) contributed to these discussions and grew out of conversations I had with fellow graduate students Terry Haywoode, Celene Krauss, and Susan Stern and was further developed in discussions with Susan Stall whom I met when I joined the faculty at Iowa State University in 1989. All five of us are white, college-educated women who were involved in local struggles and worked in multiracial coalitions on behalf of low-income and working-class communities before returning to graduate school. We hoped that our graduate training would give us the tools of analysis and research that we could bring back into these struggles for social and economic justice. Over the years, we recognized that few models of writing activist scholarship existed, especially ones that analyzed the ways women of different class, racial, ethnic, and regional backgrounds fought on behalf of their communities.

We approached our research as feminists and activists who were concerned that many of the women who fought alongside us did not define them as such. In fact, we identified closely with many other women who did not find a political home in the Women's Movement. For those of us from working-class backgrounds, the Women's Movement did not speak to us as loudly as it did to women from white, middle-class families. Therefore, a major subtext of our political and academic struggles included the desire to reconcile our community activist and feminist identities. In my experience, it has been through analyses of low-income women's political activism that I arrived at a broadened understanding of feminism that has enhanced my own political engagement. Further, I began to see that the history of the U.S. Women's Movement of the 1970s was written in a way that rendered invisible the women-centered activism of many women of color as well as working-class and poor women (see, e.g., Gluck et al., 1998).

Feminist Pedagogy and Feminist Activism

In the context of the Women's Movement of the late 1960s and early 1970s, feminist faculty across the United States developed innovative pedagogical strategies designed to draw on women students' personal experiences and to incorporate processes of dialogue and reflexivity into classroom interaction and course assignments (see, e.g., Fisher, 2001). Early Women's Studies classes used journal writing, autobiographical essays, and oral histories of family and community members among other techniques to provide students with the opportunity to explore how processes of oppression often hidden from view shaped their personal lives. The notion of the "personal is political," the central tenet of consciousness-raising groups of this period, provided the pedagogical framing for many of these classroom exercises. With the institutionalization of Women's Studies and in the context of the ongoing backlash against feminist and multicultural challenges to androcentric educational models, experiential approaches to feminist pedagogy have come under attack from critics both within and outside Women's Studies. Embedded within these diverse critiques is a fear that by valorizing women's experiences over other ways of knowing, Women's Studies classrooms do not measure up to academic standards of "excellence."

Institutional factors and conservative political forces only partially explain the challenges to experiential learning in many Women's Studies programs. Developments within feminist scholarship contributed to the increasing ambivalence toward experiential strategies in Women's Studies. Post-structuralist analyses, which call into question what "counts as experience" (Scott, 1992, 37), as well as critiques of certain essentializing interpretations of standpoint epistemologies form powerful cautions against valorizing women's experience in the Women's Studies classroom. However, feminists of color and those writing about the educational experiences of working-class women emphasized the importance of "knowledge produced through experience" (Luttrell, 1989, 40). Patricia Hill Collins (2000, 208) notes that Black women frequently use "concrete experience" as a basis from which to make knowledge claims. However, Collins points out, oppositional consciousness is not an inevitable outcome of experience (see also Harding, 1986; Hartsock, 1983; Smith, 1987).

Since 1989, I have been using an approach to experiential learning in my large introductory women's studies classes that I adapted from teaching community action in social work. My goals with the Community Action Project (CAP) were twofold: first, to teach lessons in politics, collective action, and feminist analyses; and, second, to help students develop investigatory and political skills which can complement a developing critical consciousness. From a feminist pedagogical perspective, the development of critical consciousness provides students with the ability to call into question taken-for-granted understandings of their social, political, economic, and academic life

The CAP gave students the experiential base from which to understand how to link local community actions with broader struggles for social change and feminist theorizing about these processes. As an experiential group project, the CAP also challenges the privatization of academic knowledge production. Students begin to recognize how the individualized approach to higher education inhibits the development of certain kinds of knowledge, privileging self-interest and competition among students rather than cooperation and group learning. The project also provided a context through which my students could critique the academy as a site of social control as well as discover how it is linked to other institutions to reproduce patterns of inequality. Students came to recognize their own position within these institutions as well as their ability to challenge oppressive institutional practices. In some instances, they were successful in making small changes within the university or their community.

Regardless of the specific outcome, the CAP group process offers one strategy for incorporating social-change-oriented pedagogy into the Women's Studies curriculum. In other words, the institutionalization of Women's Studies does not mean the inevitable loss of feminist praxis in the feminist classroom. Despite the contradictions of the CAP as an assignment in a required course for which a grade is given, reviewing the projects, I recognized that they continually reinspired my own sense of political possibilities for a new generation of Women's Studies students. Many of my graduate students have also used CAPs in their courses and report similar results for themselves and their students.[3]

Interdisciplinarity and Epistemological Considerations

My academic location in Women's Studies offered me the opportunity to develop a rich interdisciplinary approach to scholarship and teaching.[4] Teaching interdisciplinary feminist theory and methods has been especially challenging for me, as I suspect it is for other feminist faculty. Each discipline and interdisciplinary site has a different set of epistemological and methodological strategies that are fore grounded and influence what key questions frame the research enterprise, what counts as data or evidence, and how these data should be analyzed and presented for an academic audience. Once these issues are made visible it is possible to identify research questions that can facilitate cross-disciplinary conversations and research agendas. It is also possible to specify how each disciplinary location provides important tools to enhance interdisciplinary scholarship and multi-methodological research strategies. However, before I learned how to identify and analyze these differences, I made a number of strategic errors in my graduate courses that provided some painful lesson along the way. For example, I remember with much regret my first graduate interdisciplinary seminar in feminist methodologies where I failed to help the social science–oriented graduate students and the humanities-oriented students find common ground to mediate the larger disciplinary divide, which at a fundamental level hinged on a tension between modernist and postmodernist feminist frames.

This seminar, as painful as it was at the time, propelled me on my journey to learn how best to serve as a translator and facilitator of interdisciplinary conversations.

My strategy for negotiating the challenge of postmodernism to so-called modernism goals of feminism has been one of praxis, namely, to generate a materialist feminist theoretical approach informed by postmodern and postcolonial analyses of knowledge, power, and language that speaks to the empirical world in which my research takes place (Naples, 2003). By foregrounding the everyday world of poor women of different racial and ethnic backgrounds in both the rural and urban United States and by exploring the governing practices that shape their lives I have worked to build a class conscious and anti-racist methodological approach (see Alexander and Mohanty, 1997, xxxiii). Although acknowledging the limits of my own angle of vision and reflective practice to disrupt the power imbalances inherent in the research enterprise, my feminist praxis led me beyond the modernist/postmodernist divide to draw on some of the many valuable insights of Marxist, postcolonial, and postmodern perspectives on power and knowledge.

Linking Generations of Women's Studies Students and Feminist Activists

Feminist consciousness and political practice has transformed the minds and hearts of a new generation of activists through experiences that differ, for the most part, from the experiences that shaped feminist praxis of the previous generation of activists. Women's Studies Programs provide valuable opportunities for young activists and scholars to learn from the past and generate new theoretical and political approaches designed to contest inequality and injustice. These students come to Women's Studies with a different set of experiences and expectations.

During the later part of the twentieth century, feminists have experimented with new forms of activism. Cultural politics and cyberactivism have proven to be two of the most effective forms of organizing that contest, respectively, distinctions between culture and politics and between local and global organizing. Cultural politics expanded as a form of activism in the 1980s and 1990s as exemplified by the Guerrilla Girls and Lesbian Avengers. Lesbian Avengers (2008) drew on the lessons of ACT-UP and created spectacles such as eating fire to creatively challenge homophobia and work toward lesbian visibility. First started in 1992 in New York City, Lesbian Avengers chapters are now thriving cities across the United States and internationally.

Many of my students have been drawn to the creative performative and "in-the-streets" form of activism modeled by the Lesbian Avengers and Guerrilla Girls. The Guerrilla Girls, a group of art-activists who identify themselves through names of deceased women artists, "have been reinventing the F word—feminism" since 1985. Wearing gorilla masks in public, they have generated a series of provocative posters and "actions that expose sexism and racism in politics, the art world, film and the culture at large" (Guerilla Girls, 2008). Although much of the Guerrilla Girls' political performances are unique innovations, their use of spectacles and alternative modes of communication has precedents in the 1960s and early 1970s. For example, Hannah Wilke and Eleanor Antin and other radical feminist performance artists used spectacle, ritual, and their own bodies to challenge sexism and the patriarchal art world.

The expansion of access to media and information through cable television and the Internet has facilitated a wide range of activist efforts and has proven a significant resource for the development of feminist organizing and feminist pedagogy. Over the past decade, we have witnessed the power of cyber-activism in fighting the globalization of capitalism and the U.S. war against Afghanistan and Iraq as well as promoting domestic legislative lobbying and fundraising for diverse political causes.

My students and I are especially enthusiastic about the growth and development of transnational feminist organizing, even as we recognize that this form of activism is more difficult to enact than local community-based activism (see, e.g., Naples and Desai, 2002). The processes by which feminists and other progressive actors in different parts of the world can link their resistance strategies across boundaries remains the greatest challenge for contemporary progressive movements. Racism continues to haunt Women's Studies program as does the perception that it remains a site for white women's feminism (see, e.g., Hernández and Rehman, 2002; Sandoval, 1982).

Women's Studies has widened the frame to address queer studies, justice studies, and postcolonial studies. Some programs have grown to encompass these perspectives. Many of my students often define their interests as more aligned with these frames than with feminism. In fact, a surprising number of my students do not define their political identities as feminist, rather they see themselves as humanists interested in human rights and broad-based movements for social justice. However, these students continue to utilize a feminist lens in evaluating these organizing efforts.

Despite these challenges, I expect that feminists inside and outside of Women's Studies Programs and academia will become increasingly more sophisticated in understanding processes of racialization and globalization and will be better equipped to build coalitions across constituencies, different issues, diverse constituencies, and national borders. Remaining aware of feminism as a work-in-progress that is generated and rejuvenated through reflective dialogue and intergenerational practice will strengthen the power of feminist praxis for social justice and progressive social change and will continue to fuel the passion of Women's Studies faculty and students alike.

Conclusion

With the development of university-based Centers for Women and Gay, Lesbian, Bisexual and Transgendered students among other organizations for support and grassroots political engagement, Women's Studies classrooms are no longer the only formal sites in colleges and universities where consciousness-raising and individual empowerment occurs. We now occupy the institutional location from which to further develop strategies that link the personal to the political to the theoretical in new and creative ways. Moreover, given the growth in size of our Women's Studies classrooms (especially at the introductory level), we need to find epistemological and pedagogical approaches that lend themselves to feminist education without marginalizing the important insights of early consciousness-raising strategies. Although the institutionalization of women's studies in the academy and of women's movement goals more broadly makes it more difficult to sustain the commitment to feminist praxis within the women's studies classroom, I remain convinced of its continued vibrancy and centrality to the project of women's studies.

Women Studies Programs and Departments in some universities are expanding to incorporate graduate programs and even endowed chairs, while in other colleges and universities the programs they are being starved financially and marginalized in an increasingly corporatized environment. In the climate of backlash, we need to develop new political strategies and strengthen coalitions across interdisciplinary women's studies, sexualities studies, and ethnic studies programs and with community groups to ensure the continuity of these important academic units. Simultaneously, we need to ensure that traditional disciplines continue to incorporate feminist and antiracist perspectives and to open their doors to diverse academic practitioners.

This raises one of the last issues included in Alice Ginsberg's charge to authors in this collection; namely, what's in a name. In response to the widening of the intersectional lens and theoretical frames of feminist and queer scholarship, some programs are changing their names from Women's Studies to Women and Gender Studies, Gender and Sexualities Studies, or Feminist Studies. Other programs have been subject to administrative pressures to merge with Ethnic Studies to form Cultural Studies or American Studies programs. Although I am one of those who remain committed to the continuity of the founding frame of Women's Studies (although not as opposed to broadening the title to include gender and sexualities or feminist studies), I support the choices made by others as they respond to changes in their political, institutional, and intellectual environment.

We are living in challenging times for feminist activism and for activist scholarship. The conservative backlash against women's reproductive rights, affirmative action and immigrant rights, among other challenges to social and economic justice, is shaping the work of feminists both inside and outside the academy. From the recent appointments of John Roberts and Samuel Alito to the Supreme Court to the increase in anti-immigrant laws and curtailment of civil liberties, there is more reason than ever for community activists and feminist academics to join together to fight for progressive social change and to curb the rising conservative tide. Women's Studies will continue to provide academic homes to students searching for a way to join their activist concerns with their intellectual development and to interdisciplinary feminist scholars who refuse to be disciplined.

Notes

Portions of this essay are excerpted from Naples, 1992; 1998a; 2002; and 2005.

1. Historian Joan Kelly, anthropologist Eleanor Leacock, art historian Linda Nochlin, political scientists Marilyn Gittell and Frances Fox Piven, sociologists Judith Lorber and Gaye Tuchman all contributed to my development as a feminist in academia.
2. My thanks to Nina Fortin (d. 1987), Martha Ecker, Dawn Esposito, Terry Haywoode, Celene Krauss, Lorraine Cohen, Ros Bologh, Natalie Sokoloff, and Mary Claire Lennon who offered friendship and intellectual engagement that helped nourish me as I tried to reconcile what I considered at the time to be a retreat from activism.

3. My commitment to the development of experiential learning opportunities to enrich the educational experiences of my students led me to co-edit a book on teaching feminist activism (Naples and Bojar, 2002) designed for faculty interested in linking feminist theoretical perspectives on activism with feminist pedagogy and experiential learning.

4. Feminists in Women's Studies at the University of California were especially helpful in providing me with the support and crucial insights need to extend the interdisciplinary reach of my scholarship. My thanks to Leslie Rabine, John Smith, Jane, Newman, and Francesca Cancian who were my first mentors at Irvine. Later Robyn Weigman, Val Jenness, Kitty Calavita, Susan Greenhalgh, Sandra Harding, and Karen Brodkin offered me guidance as I struggled to balance the competing demands of a dual appointment in Sociology and Women's Studies. Through Sociologists for Women in Society, the Society for the Study of Social Problems, and other professional associations, I was fortunate to meet Myra Marx Ferree, Cornelia Butler Flora, R.W. Connell, and Dorothy Smith, all of whom continue to play important roles in my ongoing education as a feminist scholar.

References

Alexander, M.J., and C.T. Mohanty. (1997). *Feminist genealogies, colonial legacies, and democratic futures.* New York: Routledge.

Beneria, L. and S. Feldman (Eds.) (1992). *Unequal burden: Economic crises, persistent poverty, and women's work.* Boulder, CO: Westview Press.

Collins, P.H. (2000). *Black feminist thought: Knowledge, consciousness, and the politics of empowerment.* New York: Routledge.

Enloe, C. (2000). *Maneuvers: The international politics of militarizing women's lives.* Berkeley, CA: University of California Press.

Fisher, B.M. (2001). *No angel in the classroom.* Lanham, MD: Rowman and Littlefield.

Guerilla Girls. (2008). http://www.guerrillagirls.com/ accessed on January 13, 2008.

Gluck, S.B., with M. Blackwell, S. Cotrell, and K. Harper. (1998). "Whose feminism, whose history? Reflections on excavating the history of (the) U.S. women's movement(s)." In N.A. Naples (Ed.), *Community activism and feminist politics organizing across race, class, and gender.* New York: Routledge, 31–65.

Harding, S. (1986). *The science question in feminism.* Ithaca, NY: Cornell University Press.

Hartsock, N. (1983). *Money, sex and power: Toward a feminist historical materialism.* Boston, MA: Northeastern University.

Hernández, D. and B. Rehman B. (Eds.) (2002). *Colonize this! Young women of color on today's feminism.* New York: Seal Press.

Lesbian Avengers. (2008). *About the avengers.* http://geocities.com/gainesvilleavengers/aboutavengers.htm accessed January 13, 2008.

Luttrell, W. (1989). "Working-class women's ways of knowing: Effects of gender, race and class." *Sociology of Education* 62(1): 33–46.

Naples, N.A. (1992). "On becoming a feminist sociologist: A working class woman's perspective." Paper presented at the Annual Meetings of the *Midwest Sociologists for Women in Society*, Kansas City, MO, April.

———. (Ed.) (1998a). *Community activism and feminist politics: Organizing across Race, class and gender*. New York: Routledge.

———. (1998b). *Grassroots warriors: Activism mothering, community work, and the war on poverty*. New York: Routledge.

———. (2002). "The dynamics of critical pedagogy, experiential learning and feminist praxis in Women's Studies." In N. Naples and K. Bojar (Eds.), *Teaching feminist activism: Strategies from the field*. New York: Routledge, 9–21.

———. (2003). *Feminism and method: Ethnography, discourse analysis, and activist research*. New York: Routledge.

———. (2005). "Confronting the future, learning from the past: Feminist praxis in the 21st century." In J. Reger (Ed.), *Different wavelengths: Questions for feminists in the 21st Century*. New York: Routledge, 215–235.

Naples, N.A. and K. Bojar. (Eds.) (2002). *Teaching feminist activism: Strategies from the Field*. New York: Routledge.

Naples, N.A. and M. Desai. (Eds.) (2002). *Women's activism and globalization: Linking local struggles and transnational politics*. New York: Routledge.

Peavey, F. (1986). *Heart politics*. Philadelphia, PA: New Society.

Sandoval, C. (1982). *Feminism and racism: A report on the 1981 National Women's Studies Association Conference*. Oakland, CA: Center for Third World Organizing.

Scott, J.W. (1992). "Experience." In J. Butler and J. Scott (Eds.), *Feminists theorize the political*. New York: Routledge, 22–40.

Smith, D.E. (1987). *The everyday world as problematic: A feminist sociology*. Toronto, Canada: University of Toronto Press.

Stout, L. (1990). *Bridging the class divide and other lessons for grassroots organizing*. Boston, MA: Beacon Press.

12

What Year Is It?

Jean Fox O'Barr

What year is it? It's 2008 as I write, 2009 or more by the time others read this essay. I start with this question. We as feminists stand on four decades of research, teaching, campus advocacy, and public policy development. Yet in many ways, I feel like I'm living in the late 1960s. And I am not referring to fashion trends.

I begin by locating myself and my work in Women's Studies over the past 40 years and then go on to explore where I sense we are in this change process of two steps forward and one or more back. Reflections on institutionalization and the politics surrounding it form the core of my analysis. My concern is that Women's Studies maintain its political edge, for much remains to be done.

Beginnings

From the point of view of the small town where I was raised, I went to college to find a husband. Although I could not have expressed it at the time, I went in 1960 because I was a girl nerd in high school and could not imagine what else there was to do. When I graduated in 1964, change was seeping in and the Woodrow Wilson Foundation said I should be a college teacher. Continuing in school made me very happy. It also satisfied my family since I had not yet found a mate. By the time I finished my dissertation, got married and had a baby in 1969, the world had taken several fast turns—I lived with my husband-to-be before getting married, people called me doctor, fellow graduate students were criticizing their teachers as well as the government, and folks thought that my dissertation chapter on women leaders in rural

Tanzania was intriguing. Above all, discussions swirled about women and their liberation.

In 1969 I moved to Durham, North Carolina, then something of a cultural backwater and began an administrative and teaching career at Duke University where I have stayed over these same 40 years. Duke transformed itself from a solid Methodist college to a major research university. The undergraduates, while remaining the children of elites, shifted from being the children of Southerners and alumni to an international conglomerate. I have taught courses focused on women, gender, and feminism this entire time.

For 11 years (1972–1983) I headed up Continuing Education, first teaching lecture courses jointly with colleagues, courses on African women (the subject of my dissertation in Political Science) and later courses on dual careers (drawing on the experiences of the adult returning students I was advising). When I became director of Women's Studies (1983–2000), I taught the introductory courses and eventually feminist theory courses. Moving to the Program in Education after my directorship, I began teaching on gender and higher education.

I have not saved the syllabi from these courses. But I can discern a pattern that runs through my teaching (supplemented by what I saw as an administrator) that I believe characterizes the evolution of feminist scholarship and the place of Women's Studies in American university programs today. This essay is organized around those patterns. I will not for one minute argue that Duke is representative of all schools. I will only relate my experiences and suggest that there are elements in the Duke patterns that are found elsewhere.

Discovery

Duke University had had what was known as a coordinate college for women since it's inception in 1930. The Woman's College had its own administrative staff, its own residential halls and policies, its own student organizations, its own campus, and above all its own traditions. It shared the libraries, classrooms, and faculty as well as governance at the level of the Board of Trustees with the men's college, Trinity. Thus, in the late 1960s when the movement to coeducation was gaining ground, The Woman's College assessed its future and merged with Trinity. The last class graduated in 1972. One of the recommendations of the Board of Visitors that advised the college was to begin to address the entrance of women into the professions. Implementing that recommendation—moving the women of

Duke University from what was institutional marginalization to the center—brought both costs and benefits and serves as a harbinger of what would happen to Women's Studies some 30 years later.

In 1968, on the cusp of consolidation, three faculty—from Psychology, Economics, and History—came together to offer a half-credit course called "The American Woman: History and Prospects." When I began as an administrator in 1972, I joined the loose coalition of faculty offering the course. At that time, we filled the Chemistry auditorium—356 seats—with men and women interested in the women's liberation movement as it was then called. Guest speakers came each week to address a variety of topics from health care to literature. There was no payment for faculty or guests. We were fired up, certain that our discoveries would change the world.

And to some extent they did. On the student level, individual students dove into the possibilities. They raised women-centered questions in their other classes. I remember one professor specializing in Asian history who told me that the questions raised by one of the women in his class had caused him to work up lectures emerging from the new scholarship in his field and even put a question on women's roles on the final examination. He, of course, did not know that she had come to me and we had plotted out a series of questions over the semester that caused him to undertake new work. She and I took delight in our small and still secret intervention for years.

Students organized groups to discuss the issues, groups that grew and declined as individual leaders became involved and then quickly graduated. Most of all, a number of those individual students became intrigued with possibilities of the emergent fields and went on to graduate school. Some were in the first cohort of graduate women to specialize in feminist scholarship and later others were in the first classes of the programs offering the Ph.D. in Women's Studies. A number now direct or teach in Women's Studies programs elsewhere. Still others continued in law, medicine, and social work, all with an emphasis on women and policy.

At the level of the faculty our discovery was equally exhilarating, reinforced by the sharing we did. Everyone had a copy of Vivian Gornick and Barbara K. Moran, *Woman in Sexist Society: Studies in Power and Powerlessness* (Signet, 1971), Robin Morgan, *Sisterhood Is Powerful: An Anthology of Writings from the Women's Liberation Movement* (Vintage, 1970), and Michelle Rosaldo and Louise Lamphere, *Woman, Culture and Society* (Stanford, 1974). We read the articles and debated them. Wasn't Linda Nochlin's claim applicable to our own field? Why had no one ever formulated child development like Nancy Chodorow? How could we not have seen what Pat Mainardi saw?

We organized a research conference for faculty, students, and community people, modeled on professional conferences. Students moderated the panels and served as discussants to papers written by combinations of faculty and students. Called "The Changing Roles of Women and Men" in 1975 it was the first such conference on this campus.

Discovery was in the air and everyone believed we were going to move rapidly to a new academy. And we felt supported by the students, encouraged by a few female faculty hires (no one said feminist scholar yet), and were blissfully unaware of the opposition if there was any—we were too excited about the discovery and naively confident that everyone else would be so as well.

Involvement

Founded officially in 1983 Women's Studies at Duke seemed to soar in the 1980s. Faculty participation was at an all-time high, seen in monthly seminars. The curriculum was expanding, eventually resulting in a major. Along with scholars from neighboring universities, the editorial offices of *SIGNS: Journal of Women in Culture and Society* were housed here. And university sanction seemed evident in our fundraising success. They were heady times. The burnout and oddity of our previous informal endeavors gave way to an officially designated program and a sense of belonging to the mainstream. In retrospect, I believe we felt a balance between challenging and belonging to the institution.

The growing number of faculty interested in the study of women and gender, women and men eagerly involved themselves in the program. We began a series of monthly dinners at which one faculty member shared their work with an interdisciplinary audience in mind. We heard about feminist biblical studies, new research in the cognitive sciences, and how cafeteria benefit programs both helped and hindered female professionals. Before and after the simple buffet supper, faculty talked. They talked about professional issues (research plans, conference offerings, new courses, tenure procedures), social concerns (pending legislation, local women's centers, presidential politics) and personal questions (do you play tennis? How about a movie Saturday? Are you happy with your day care?). The numbers varied from a dozen to three and every session provided sets of connections that contributed to both individual and institutional successes.

The format was so successful that a few years later, the Women's Studies program began to offer a separate dinner series for graduate students. At one point in the late 1980s, more than a hundred graduate students were involved. Sometimes students presented or practiced conference presentations; other times we brought visiting scholars to them to discuss new work. The decision to maintain two groups was deliberate: each group enjoyed the freedom of solidarity in these dinners. And they had ample opportunity to mix in other formats.

The curriculum exploded. Courses were offered by the program as well as by faculty affiliated with the program. Departments seemed happy to have these enrollments and rarely seemed interested enough in their content to rein in faculty choices. With a growing course selection, we had a growing number of minors and eventually majors. The excitement was not just in the number of students taking courses but also in the intellectual involvement that arose from students taking more than one course, taking courses together, and finding that they could walk their talk in a variety of student projects.

One example: when a group of fired-up students came to Women's Studies and asked for conference funding—"If we just have a really good conference, I know Student Affairs will set up a Women's Center" is the direct quote—I said no. My experience told me they were inexperienced in the complexities of campus politics. My counteroffer was to show them how. And we did. Seven students took an independent study with me, figuring out best practices on other campuses, interviewing everyone in student affairs for their understanding of the needs of female students, conducting a campus survey, coordinating with student groups, and designing a model program. One semester, one report, and one faculty taskforce that blessed their lengthy report (duplicated at night on the Women's Studies copier since the project had no funds) and the Women's Center was a reality in 1987.

Amidst all this campus activity, a national meteor struck. Duke had a cooperative research center with UNC-Chapel Hill; the faculty associated with it were asked to develop a proposal to be the next editorial offices for *SIGNS: Journal of Women in Culture and Society*. We decided to plunge ahead, thinking that we would practice and be awarded the journal in the next five-year cycle. One hot July day in 1984, I got the word: I had been selected to be editor, hire two staff and set up the offices at Duke. In that initial whirlwind of activity, I remember best a memo from a faculty member. Written by hand on those old yellow half-sheet memo pads, he told me that a prospective graduate student they were eagerly courting had decided on Duke because *SIGNS* was here. Could I tell him what she meant?

The Women's Studies faculty eagerly embraced the arrival of *SIGNS*. Along with regional colleagues, they sat on the editorial board, they read manuscripts, they offered ideas. In short, our national involvement was just another step in the growth of feminist scholarship locally.

One additional group became involved in Women's Studies, adding their resources to those of the campus. Duke had female alumna who had never quite reconciled their positive experiences in The Woman's College with the changes of the 1970s and 1980s. Women's Studies began a newsletter and a series of educational weekend programs late in the 1980s. In two decades of activities, some 1,500 of alumna became involved. They attended programs, they spoke on behalf of Women's Studies in their other university endeavors, and they provided resources for our efforts in the form of endowments. In fact, Duke's first female president, Nannerl Keohane, made her initial visit to Duke for a Women's Studies Alumna Weekend.

Duke was growing into a national research university. Women's Studies seemed a key part of that growth—outstanding faculty, strong students, alumni approval, peer recognition, and funds to carry it all out.

Institutionalization

The third decade brought what many of us in Women's Studies saw as a dream state: large enrollments (at the graduate level as well), substantial endowments, extensive programming, plans for internships, graduate degrees, and continued cooperation with numerous other university programs. We began by securing a major in the early 1990s and began readying for our final goal: departmental status and tenure lines. Ellen Messer Davidow had not yet published *Disciplining Feminism: From Social Activism to Academic Discourse* (Duke University Press, 2002) or we might have had a guide to what lay before us.

Each campus offers its own complex set of policies and procedures, some written, some custom, often contested if not modified, for achieving departmental status. Ours was no exception. We began with the assumption that what we had done to get a major would work again. At that time, we had followed classic lobbying techniques. We secured the endorsement of the dean, already predisposed to Women's Studies. We took the list of members of the decision-making council, assessed their positions, reinforced the supporters through individual contacts, ignored the hard-core opposition and lobbied those in between. My favorite antidote was the decision of one colleague to help her neighbor rake leaves and secure his support in the course of their activity. On the day of the meeting, several Women's Studies faculty attended and sat next to the wavering members—we believe we neutralized at least three votes when those men abstained.

When it came to departmental status, we were in what Southerners call "high cotton," more difficulty territory. Although we had administrative support, the decision-making body was smaller, closed, and peopled by colleagues who were less sympathetic. Old rivalries, some intellectual, some personal, arose. The proposal was turned down on the first round. With the assistance of the dean and the recruitment of some senior advocates, it passed on the second. The formal decision was in; now the politics of implementation began.

The staff and faculty of the program breathed a sign of relief—prematurely it turned out, for as another old saying goes, "the devil is in the details." Now we imagined we could get on with the work at hand: reorganizing the program structure, developing plans to recruit faculty, and envisioning the next decade. During the several years it took to propose and achieve departmental status, it is important to note that all of the other program components were humming along. The number of undergraduate majors, particularly the number doing honors, grew. Graduate students from across the university sought the program out and engaged in teaching and research assistantships, got travel awards, taught seminars, and generally found an interdisciplinary intellectual home that enhanced their departmental training. Faculty willingly participated when speakers came, when committees were needed, when advice and support were sought. Collaboration with other university programs was frequent and usually smooth. And alumnae support kept increasing—monies for teaching and research, symposia funds, even an endowed chair for the to-be program.

In retrospect, I understand that the buzz within Women's Studies differed from the evolving buzz in the university as a whole. None of us did at the time. I have come to understand that Women's Studies in the late 1990s was viewed by the administration and some faculty as too political. By political, I mean that we talked in terms of who had power, how they used it, what that meant for women on campus, and if and how we might influence power dynamics for the benefit of women. The program's organizational structure was open. We had a director, an associate director, three staff members, and a series of faculty-student committees. More importantly, we had a welcoming meeting-lounge-lunchroom with mailboxes, machines, and a kitchen where everyone (including the maintenance man who was a labor organizer) was welcome to come and talk. Ideas and issues flowed from the constant gatherings in that backroom. Individual counsel, small group projects, larger initiatives all flowed from the conversations that began in this atmosphere that enabled combining analysis and action.

Outside our offices, the university was changing. It was moving to the forefront of research universities worldwide. International reputations were rewarded more than campus leadership. Specialization in teaching and research became more valued than general knowledge. Innovations trumped traditions. Many faculty focused on their careers and found less relevance in an interdisciplinary program. Doing anything in a nonstandard way was discouraged. Challenging these new power relationships became risky. Women's Studies faced a classic dilemma—the choice between going along to get what we wanted and resisting the process being laid on us. Over a two-year period, the administration undertook the process of reconstituting the program, appointing a committee primarily outside the program to make decisions about directions and hiring. My role as founding director was coming to a close after two decades. I anticipated retirement; my own academic status as a non-tenure-track professor meant that I was not in a position to participate in this next phase, a fact I clearly understood and supported from the beginning of the process.

The change can be seen most explicitly in the fate of the Alumna Council. It was disbanded shortly after the department was set up in 2000. The decision came from the dean and was delivered by the president. The stated reason was that Women's Studies was now a full academic unit, supported by the university; in addition, no other department had such a body. Beneath the public reason was a complex political web: tensions between the new director and the Council; an administrative desire to make Women's Studies like every other department; the presence in many departments of feminist scholars who no longer needed the interdisciplinary intellectual home of Women's Studies; the coherence that once characterized work on women and gender was enriched, complicated, and challenged by an increased emphasis on race and sexuality.

Departmentalization

It is 2008 as I write. This year, like every year, I receive Christmas letters in the hundreds, from former students over a 40-year teaching span. Almost a dozen are directors of Women's Studies and another 50 are faculty members, telling me of the breadth and depth of what they are doing, saying that the Women's Studies program they knew during their Duke years continues to inspire them. They tell me of political successes, ongoing challenges, and new adventures. Larger numbers of former students and alumni from the former Council report that they continuing as feminists—as nonprofit managers, parents, teachers, business owners, attorneys, doctors, and ministers. These communications hearten me because I read in each one an ongoing commitment to changing the gender dynamics of family, work, and community, here and abroad, on the twenty-first century terms.

It is also January as I write. The new semester begins in a week. Preparing the upcoming syllabi, I am struck by the contrast I see between what I have read in the holiday communications and who I will meet around the seminar room table next week. The students I will teach can tell me about the civil rights movement but not the women's movement; they know almost no women's history; they cannot infuse gender into discussions of contemporary policy issues; nor are they able to see power dynamics in social processes. They will tell me that they are set individually because, as their logic goes, institutions will change because they are qualified. Most will insist that they pay no attention at all to the media cultural stereotypes and neither do their families or friends. Clearly, some of these attitudes are the result of teaching students in a highly selective university—but there is more to it than that, I believe.

In one sense, next week's students do not differ from those at the beginning of any semester over the past four decades. My purpose in teaching is to offer students new knowledge, to help them integrate those perspectives into their previous ones, and to suggest ways of readjusting their gender lenses. But I find my work in this decade more difficult than my work in the previous decades. Part of that difficulty stems from the culture at large. But that culture has always been a difficult, albeit in other ways.

I attribute part of my difficulty to the lack of a feminist consciousness in the classrooms and student spaces around me. The campus culture has shifted to value concepts over conversation, individual prestige over community. The faculty is more depoliticized, looking out for themselves. The students are more career-minded, looking out for themselves. The administration is more financially minded, wavering about structural models and often choosing the corporate one. The voice of feminist scholarship comes across as weak in this din. It is true that alternatives have arisen. There is a women's leadership program. The Women's Center in Student Affairs charges on. The library has an outstanding women's archives collection and programs to bolster their use. Faculty get grants for research that incorporates women. Women athletes proliferate. Nonetheless, the scholarship and its partner, teaching, that undergirds, sustains, and advances these endeavors are missing. Students are not helped to link feminist knowledge with action. Faculty are not situated so that they are encouraged to think about their work in relation to politics and power.

How do I understand what I see? Women's Studies was once a marginalized place where people had to make connections—across disciplines, from gown to town, between daily endeavors and intellectual questions. Making connections is inherently political and politicizing work; this work suffers with the success of departmentalization unless it is given specific attention. With all the accoutrements of legitimacy (space, lines, budgets, leaves, etc.), little attention goes to creating the space to challenge the very privileges the department now enjoys.

I am reminded of the late 1960s when women's colleges were considering coeducation and the dominant voices said women are equal now, so let's move on. There were few feminist scholars around then to analyze what the move would mean and what correctives and compensations might be needed. Women's Studies arose at that very time; it created a body of knowledge, an approach to teaching, a power-based analysis of structures and systems. A community grew to demand change. In many senses, those advocates got the change they wanted. Now, some 20, 30 years later, new questions arise. Do we need another cycle of change? If so, who will instigate it? What will its focus be? Where will we gather to explore it? Gerda Lerner, the preeminent women's historian argues that change in America always comes from the bottom up; she believes that such change is percolating now. I want to believe her and will start class with that premise, believing that the labyrinth to equality would be easier to discern if Women's Studies can achieve a balance between its newfound and well-deserved institutional status and its history of marginality and challenge.

Note

I am grateful to Ann Marie Rasmussen and Erin Smith for comments on an earlier draft and to Gerda Lerner for ongoing conversations.

About the Contributors

Evelyn Torton Beck is Women's Studies Professor Emerita at the University of Maryland and an Alum Research Fellow with the Creative Longevity and Wisdom Initiative at the Fielding Graduate Institute from where she completed a (second) Ph.D. in Clinical Psychology (2004). She is the editor of *Nice Jewish Girls: A Lesbian Anthology* (1982/1987) and has published widely in a variety of interdisciplinary fields. Currently, she is reworking her dissertation ("Physical Illness, Psychological Woundedness and the Healing Power of Art in the Life and Work of Franz Kafka and Frida Kahlo") for publication and preparing a collection of her essays. When not writing, she teaches Sacred Circle Dancing (about which she is passionate) at community institutes of life-long learning in the Washington, DC area. She also offers interarts workshops on diverse themes: Creative Aging, The Stages of a Woman's Life, Health and Healing.

Margaret Smith Crocco is Professor and Coordinator of the Program in Social Studies at Teachers College, Columbia University, where she has worked since 1993. Before that, she taught high school in New Jersey and at colleges in Maryland, Texas, and New Jersey. Her professional life and scholarship have focused on issues of gender equity, diversity, schooling, and social studies.

C. Alejandra Elenes is an Associate Professor of Women's Studies, and Affiliated Faculty in the Ethnicity, Race, and First Nation Studies Programs at Arizona State University. Her research centers on the application of borderland theories to study the relationship between Chicana culture and knowledge, and how it relates to pedagogy and epistemology. She has published numerous articles and chapters in books, and is coeditor of the book *Chicana/Latina Education in Everyday Life: Feminista Perspectives on Pedagogy and Epistemology* (2006) published by SUNY Press (winner of the 2006 American Educational Studies Association (AESA) book critics' award). She is also coeditor of the special issue "Chicana/Mexicana

Pedagogies: Consejos, Respeto y Educación in Everyday Life" for the *International Journal of Qualitative Studies in Education* (2001). Currently, she is working on a book manuscript, *Transforming Borders: Chicana/o Popular Culture and Pedagogy.*

Alice E. Ginsberg spent eight years as a Program Officer at the Pennsylvania Humanities Council, where she developed and directed women's studies programs across the state. These included GATE (Gender Awareness through Education), a three-year professional development program in Philadelphia urban public schools. She holds a Ph.D. in Education from the University of Pennsylvania and is the coauthor of *Gender in Urban Education: Strategies for Student Achievement* (Heinemann, 2004) and coeditor of *Gender and Educational Philanthropy: New Perspectives on Funding, Collaboration and Assessment* (Palgrave, 2007).

Judith Lorber is Professor Emerita of Sociology and Women's Studies at Brooklyn College and the Graduate School, City University of New York, where she was the first Woman's Studies Coordinator. She is the author of *Breaking the Bowls: Degendering and Feminist Change, Gender Inequality: Feminist Theories and Politics, Paradoxes of Gender,* and *Women Physicians: Careers, Status and Power.* With Lisa Jean Moore, she is the author of *Gendered Bodies: Feminist Perspectives,* and *Gender and the Social Construction of Illness.* She is coeditor of the *Handbook of Gender and Women's Studies* (with Kathy Davis and Mary Evans), *Revisioning Gender* (with Myra Marx Ferree and Beth B. Hess) and *The Social Construction of Gender* (with Susan A. Farrell). She is the Founding Editor of *Gender & Society,* official publication of Sociologists for Women in Society.

Jean Fox O'Barr, a political scientist by background, served as the Director of Women's Studies at Duke from 1983 to 2000. During that time, she headed up the editorial board for *SIGNS: Journal of Women in Culture and Society* (1985–1990). She has edited and written a number of books, including *Feminism in Action: Building Institutions and Community through Women's Studies.*

Nancy A. Naples is Professor of Sociology and Women's Studies at the University of Connecticut where she teaches courses on gender, politics, and the state; sexual citizenship; feminist theory; feminist methodologies; feminist pedagogy, and women's activism and glo-balization. She is author of *Feminism and Method: Ethnography, Discourse Analysis, and Feminist Research* (Routledge, 2003); *Grassroots Warriors: Activist Mothering. Community Work, and the War on Poverty* (Routledge, 1998); and *Community Activism* *and Feminist Politics: Organizing Across Race, Class, and Gender* (Routledge, 1998). She is also coeditor of *Women's Activism and Globalization: Linking Local Struggles with Transnational Politics* (with Manisha Desai, Routledge, 2002) and *Teaching Feminist Activism* (with Karen Bojar, Routledge, 2002).

Sue V. Rosser serves as Dean of Ivan Allen College, the liberal arts college at Georgia Institute of Technology, where she is also Professor of Public Policy and of History, Technology, and Society. She holds the endowed Ivan Allen Dean's Chair of Liberal Arts and Technology. From 1995–1999, she was Director for the Center for Women's Studies and Gender Research and Professor of Anthropology at the University of Florida-Gainesville. In 1995, she was Senior Program Officer for Women's Programs at the National Science Foundation. From 1986 to 1995 she served as Director of Women's Studies at the University of South Carolina, where she also was a Professor of Family and Preventive Medicine in the Medical School. She is the author of ten books, including *Female-Friendly Science* (1990) from Pergamon Press. Her latest book *Women, Gender, and Technology* (2006) is co-edited with Mary Frank Fox and Deborah Johnson.

Paula Rothenberg lectures and consults on curriculum transformation and issues of inequality and privilege at colleges and universities around the country. Her articles and essays appear in journals and anthologies across the disciplines and many have been widely reprinted. Formerly Professor of Philosophy and Women's Studies at William Paterson University, in Wayne NJ, and Director of The New Jersey Project on Inclusive Scholarship, Curriculum and Teaching, she is currently Senior Fellow at the Murphy Institute, CUNY. Author of *Invisible Privilege A Memoir About Race, Class, and Gender* (University Press of Kansas, 2000, paper 2004), her widely used diversity text *Race, Class, and Gender in the United States* (St. Martin's/Worth) is now in its seventh edition, and her text anthology *White Privilege: Essential Readings on the Other Side of Racism* (Worth, 2007) is now in its third edition. Her newest college text *Beyond Borders: Thinking Critically about Global Issues* was published in 2005.

Ann Russo is an Associate Professor and currently the Director of the Women's and Gender Studies Program at DePaul University. Her research, teaching, and activism over the past 25 years have been embedded in the social movements organized to address the pervasive sexual, racial and homophobic harassment, abuse, and violence in women's lives. She is the author of *Taking Back Our Lives: A Call to Action in the Feminist Movement* (Routledge, 2001); coauthor of *Pornography:*

The Production and Consumption of Inequality (Routledge, 1998); coeditor of *Talking Back and Acting Out: Women Negotiating the Media Across Cultures* (Peter Lang, 2002); and *Third World Women and the Politics of Feminism* (Indiana University, 1991).

Joan Poliner Shapiro is Professor of Educational Administration in the Department of Educational Leadership and Policy Studies at Temple University's College of Education. Previously, at Temple, she served as an Associate Dean for research and development and as a chair of her department. She also has been the codirector of the Women's Studies Program and a supervisor of intern teachers at the University of Pennsylvania. She has coauthored the books, *Reframing Diversity in Education* (Rowman & Littlefield, 2002, paperback), *Gender in Urban Education: Strategies for Student Achievement* (Heinemann, 2004), *Ethical Leadership and Decision Making in Education: Applying Theoretical Perspectives to Complex* Dilemmas (Erlbaum, 2005, 2nd ed.) and *Ethical Educational Leadership in Turbulent Times: (Re)Solving Moral Dilemmas* (Erlbaum, 2008).

Beverly Guy-Sheftall is Professor of Women's Studies, Emory University, and is founding director and the Anna Julia Cooper Professor of Women's Studies at Spelman College in Atlanta, Georgia. Her most recent publication is *Gender Talk: the Struggle for Women's Equality in African American Communities* (Random House, 2005), which she wrote with Johnnetta Betsch Cole. She has also been the founding coeditor of *Sage: A Scholarly Journal of Black Women*, which is devoted exclusively to the experiences of women of African descent.

Jill McLean Taylor is a Professor of Education and Women's and Gender Studies at Simmons College, and is chair of Women and Gender Studies. She is coauthor with Carol Gilligan and Amy Sullivan of *Between Voice and Silence: Women and Girls, Race and Relationship*, and coeditor with Carol Gilligan and Janie Ward of *Mapping the Moral Domain: A Contribution of Women's Thinking to Psychological theory and Education*. She is also the author of many articles on girls' development and continues to conduct research on adolescent girls as well as single mothers who are receiving social welfare benefits and attempting to complete postsecondary degrees. Dr. Taylor has presented her work at many international conferences in addition to those in the United States and has most recently presented work with Latina girls who were part of GEAR UP, a federally funded, seven-year partnership between Simmons College, Suffolk University, and the Boston Public Schools.

Index